NOW, FLY WITHOUT FEAR

(Includes Aviation Glossary and Index)

By
Jim Remington, Ph. D.
Leona Remington

Inner Marker To Growth
St. Louis, Missouri

Cover Design and Illustrations: Pattie Grissom

Publisher's Cataloging in Publication

Remington, Jim, 1929-
Avioanxiety becomes controlled: now, fly without fear/Jim Remington; Leona Remington.
p. cm.
Includes bibliographical references and index.

1. Fear of flying. I. Remington, Leona, 1944- II. Title. III. Title: Fly without fear.

RC1090.R4 1992 155.965
QBI92-617

ISBN 1-879855-01-1

Printed in the United States of America

c.1

TO:

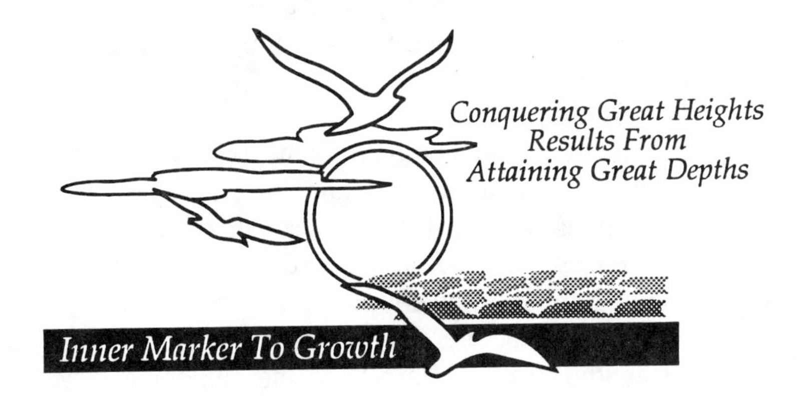

"With wings I have
won for myself
In fervent love
I shall soar .."

Gustav Mahler

DISCLAIMER: Prior to marketing, this program was specifically designed for and tested on people who experience exclusive anxiety and fear about flying on commercial jet airliners. This program is not a therapy tool for either extreme phobias or a wide, generalized range of anxieties and fears. This program does not deal with medical issues.

Most human beings are a combination of complexities with influences from the conscious, subconscious, and unconscious. There is no possible way that all of these factors and influences may be anticipated in any one particular individual by the developers of this program. Thus, the exclusive intended use of this program is to help each person using the program to reduce the anxiety of flying on commercial jet airliners.

ACKNOWLEDGMENTS

When I was eight years old, an airplane literally flew across the horizon of my consciousness. A love affair began right then. The thought crossed my mind—and has been repeatedly confirmed since then—that God was blessing my life in a very special way. I began to sense that my life was developing within a Divine Plan. The Supreme Being continues to bless my life with my involvement in the whole thrilling world of aviation.

The Divine Blessing in my life does not end there. People—significant people—come into the horizon of my consciousness to bless me. This book that will help thousands of people to fly on a commercial jet airliner with a reduction of fear and anxiety and with an increased enjoyment of flying, could not have come into existence without at least the following persons: Catherine Collins, Marketing Consultant; Dorothy Davison, Proofreader; Sheila Walker, Copy Editor; Leona Remington, Co-Developer; photo credit to Knoblauch Photography, St. Charles, Missouri.

To the brave and patient people who taught me to fly and to the thorough and patient people of the FAA who granted me pilot licenses and aircraft ratings, I am indebted. But especially to my mother who far exceeded the patience of Job tolerating my room being in constant turmoil and disarray with my model airplanes and aviation books, I am eternally grateful.

To my mom, my Higher Power, and to all the people who have blessed and continue to bless my life, I dedicate this book. It is so!

Jim Remington

CONTENTS

CONTENTS

1

ON USING THIS PROGRAM

LET'S GET ACQUAINTED

This is Jim Remington with a personal welcome to you who have purchased this exciting program. Thanks for attending to the preceding sections of Disclaimer, Copyright, Acknowledgments, and Contents.

I wish for you to understand that this is an educational book/program. The book/program is educational, and with therapeutic value.

The overall thrust of this complete book/program is an education into the wide scope of commercial aviation. You will receive, or confirm what you already know, a tremendous amount of information regarding the jet airliner, flying on a jet airliner, and airports. Some users of this book/program elect to read the glossary before beginning the actual therapeutic program.

The therapeutic value is the design and intention that this book/program will assist you in reducing your anxiety and fear about flying and to increasing your enjoyment of flying.

The intent of the authors is to meet you where you are, and assist you as you may need that assistance.

You, your real, basic need, and us meeting that need determined what is included in this book/program. We have tried very hard to integrate a "personal quality" in this education program—coming from the personal us to the personal you.

You will find in this book/program what is relevant and helpful, the professional combined with the personal.

As you move through the contents, you almost certainly will become aware of the unique and effective approaches to reduce anxiety and fear that distinguish this program from others. One of the distinguishing characteristics of this program is that the user moves in a consistent and sequential way through the material and information. One item leads into the next; that item builds on previous items. There is a smooth and common sense flow of learning and of growing.

I wish for you to "meet me." That's me, on page 11. That picture is me, holding an object—an object of my interest and involvement for many years—an airplane.

My hope in this initial contact with you is that we will establish a valuable relationship. "I greet you, and I welcome you."

As you will note when you begin reading the material, I incorporate approaches in this program that combine the mind, body, and spirit. The program benefits from my 45 years of flying and is exclusively devoted to the reduction of your anxiety about flying and to the increase of your enjoyment of flying. The program, and I as its main developer, are dedicated to you flying relaxed and confident and feeling secure, yes, to you soaring like the eagles.

Throughout the program you will sense a lot of feeling and caring. Much of these good feelings are from my co-developer, my friend, my wife, my soul-mate, Leona. The picture of the female passenger on page 67, touching the airplane—blessing it—as she is entering the airplane is Le.

Now, please, I invite you to do an imagination exercise: image yourself at a busy airport, say, St. Louis, Chicago, Washington D.C., Miami, Atlanta, Los Angeles, Denver. Here at St. Louis, in my imagination, I "see" two, parallel main runways. Commercial jet airliners are literally landing and taking off on each runway, about every one and one half minutes.

Imagine an airport or airports of your choice. Got one or more in your "mind's eye?" So many—countless—warm welcomes and loving embraces will be given and received; increased business contacts will be made; exciting places and thrilling scenes will be experienced—all because of people flying on commercial jet airliners. Hundreds and hundreds of people, pretty much like you, are and will be arriving and departing just during the few minutes you are reading this one section. Thousands and thousands of miles will be traveled in this same short time. How rewarding and exciting is flying on commercial jet airliners!

Now I get personal with you about me: I love flying for a variety of reasons. The invention of the marvelous mechanical wonder—the airplane—and the fantastic developments in airplanes and in the whole field of aviation and flying have been rich blessings to me.

A great many years ago I realized that it was important to me that other people also enjoy flying. That desire has persisted over the years, and that commitment finds expression in this rewarding program in which you are involved.

There's another picture of me on page 66. I am welcoming you aboard a commercial jet airliner. I image you walking towards me to your seat, sitting down, buckling your lap belt, and continuing to work this program designed with you in mind.

I image you on a short-focus basis successfully completely each phase of the flight; I image you on a long-focus basis getting out of your seat, walking out of the airplane, up the jetway, and into the terminal building at your destination: you have successfully reduced your anxiety about flying by actually making a trip by airplane. I develop this affirmation for you: I AFFIRM YOU MAKING THIS AIRPLANE TRIP WITH A REDUCED ANXIETY ABOUT FLYING AND WITH AN INCREASED ENJOYMENT OF FLYING.

Let's confront three important matters: identification of your fear or fears; possible "hidden" reasons causing your anxiety about flying; a computation of the practical values of traveling by commercial jet airliners.

Fear is almost always, if not always, lessened when we identify and accept our fear or fears. Many airline passengers who are anxious about flying, discover their main fear or fears in one or more of these: fear of heights, fear of loss of control, fear of crashing, fear of being in a confined area [claustrophobia].

Lesser but contributing fears may include fear or anxiety when the engine sound changes, especially when the engine sound decreases; fear of being in a storm, especially that the airplane will be hit by lightning; fear when the airplane changes positions, in particular when the airplane's nose is raised or lowered, and/or when the airplane rolls [banks] to one side or the other, especially if you are sitting on the downside of the roll; fear or anxiety when flying in dense clouds or at night, with an especial concern when landing.

I invite you to an experience of self-examination. Attempt to identify the fear or fears, the anxiety or anxieties, that you have. The next step after this identification will be to accept those as your existing fears and anxieties at this moment.

I suggest that you turn to the back of this book, beginning with page 156, and on those pages list all the fears and anxieties that you can identify. Use as many pages as are needed. Please do this now.

Now that you have before you on the sheet or sheets of paper a list of your fears and anxieties, do this: write in BOLD letters across the whole list, touching every fear and every anxiety, these words: "I SHALL REDUCE YOU. IN TIME, I SHALL OVERCOME YOU. I SHALL RECLAIM AND THEN POSSESS POWER AND CONTROL OVER YOU!" Please do this now. Focus for a time on the truth that you have proclaimed. Allow that truth to become a part of you.

When you have accepted this projected truth, please return to this section and to this point.

Upon completing this self-examination and accepting the truth involved, you may well be experiencing a sense of self-power, of self-confidence, the thrill of anticipated accomplishments. For a moment [as long as you wish], stop reading, and allow this confidence-building sensation and truth to flow into you.

Next, continue some self-analysis and try to discover so called hidden reasons that may be causing you to either feel anxious about flying or just outright cause you to have a fear about flying.

All of the following examples were real to persons who used this program and who successfully completed the program and thus reduced or eliminated their anxiety or fear about flying. You may feel a "kinship" with the people as you share in their experiences, perhaps even saying to yourself, "I can relate to that. I know that feeling. I've been, or am now, there!" One woman discovered that she really did not want to visit her mother who lived away a considerable distance. Every visit became an emotional nightmare and one big and long negative experience for the woman. Slowly she discovered that she had a hidden cause of her fear about flying: one part of her cried out that she did not want to visit her mother; the other part of her demanded that she visit her mother. After all, hammered this part, "Every daughter ought to want to visit her mother!" And a big hunk of guilt was ready for her to dump onto herself. This ambivalent feeling—this intense emotional tug of war—was "solved" with the origin of her fear about flying: "If I am afraid to fly, and driving by car or traveling by bus or train that long distance is totally impractical, then I don't have to feel guilty about not visiting my mother. It is my fear, not me, that is to blame."

Another person's first flight was in a small airplane. During this flight the pilot made abrupt maneuvers, with no advanced warning and with no explanation to the passengers, and frightened the person about flying from then until she started this program.

Still, another individual was caught in a severe storm while flying, and from that point on he was nearly petrified about flying in any airplane in a storm.

A woman was forced to make a trip by air that meant going into a situation that was very distasteful to her, so that person came to associate only negative experiences with flying.

What many time happens, however, is that a person has flown many times, over a period of years, with minimum or even no anxiety. Suddenly there develops a severe anxiety or fear about flying, and no precipitating event can be identified. Some people are anxious, or fearful about flying, even though they have never been in a commercial jet airliner. The cause of their fear is a fear of the unknown, and/or fear of change.

Did you feel a kinship, an emotional relationship, with one or several of these persons and their experiences? Probably so.

Move now to focus on you, and spend some time seeking any possible hidden cause of your anxiety or fear about flying. Remember: all of the examples just presented were real stories of real people who worked this program and overcame their anxiety. Their anxiety about flying became controlled. Also, if you draw a blank in looking for any hidden cause in you, that's no big deal. Simply move on.

We go to an interesting and potentially helpful part—to the practicality and desirability of flying by commercial jet airliners when compared with other forms of transportation. This can be a fun, interesting, and helpful activity if you choose to do it.

Turn to Appendix B, page 114. Notice, first, at the top of the first page the word "TRIP." Fill in with "from—to—return." Be sure that your computations are based on a round trip.

All of the vertical columns on the pages contain various forms—seven to be exact—of the major types of transportation. The items on the left side of the sheets are the specific subject areas to be computed and evaluated. Note that several pages are all part of this activity. Please begin working that section now.

When you have filled in the appropriate blanks and columns and have come to the last sheet of that activity, look carefully at the bottom of the page at the "Overall Final Comparison." Do the next step "Final Comparison Ranking." End with doing the "Final Comparison Ranking."

As you study the very last item, "Final Comparison Ranking," you will no doubt arrive at a very conclusive—and PRACTICAL—reason for deciding to fly by a commercial jet airliner over other forms of transportation, especially for long trips, and/or when time is limited, and/or when your desire to be somewhere or with someone is very high.

A VERY BASIC RIGHT

One of our most basic rights is to share only what we deliberately choose to share.

You possess the right to share none, some, or all, of your anxiety or fear about flying with any one else.

We fully support you in this right. Your right to keep the kind and purpose of this program protected is given to you. We underscore this very basic right and dignify you in the process of exercising the right.

You immediately will sense a truth in a theme that will be repeated throughout this program: you can take control, a control over what you may have thought to be out of your control. Assuming control is an integral part of growth. You are in control of your sharing—that is your basic right.

Pause for a moment to deepen this truth in yourself by repeating this affirmation:

> I POSSESS THE RIGHT TO SHARE ONLY WHAT I CHOOSE TO SHARE ABOUT ME WORKING THIS PROGRAM WITH ANY OTHER PERSON.

That is so!

Finally, go to Chapter 2, page 11, "The Developer of

this Program," and begin reading the program, a program designed with you and your need in mind.

With Chapter 2, move through Chapters 3, 4, 5 and 6 of this rewarding program, a movement that will make all your efforts more than worthwhile. I'll be with you in spirit and in writing.

Be assured, I am with you all during your working of this fantastic program, especially when you are working all the phases and making your actual airplane flight.

2

THE DEVELOPER OF THIS PROGRAM

This is Jim

Dr. Jim Remington

Dr. Remington holds degrees and has extensive experience in three areas that relate directly to avioanxiety: adult education, counseling psychology, and practical theology.

This program developer holds the following degrees: B.A. in psychology from Macalester College, St. Paul, Minnesota; B.D. (Masters) in Practical Theology (Counseling) from The American Baptist Seminary of The

West, Berkeley, California; S.T.M. (Masters) in Counseling from Andover Newton Theological School, Newton Centre, Massachusetts; M.A. in Guidance and Counseling, Saint Louis University, St. Louis, Missouri; Ph.D. in Adult Education Program Development, Southern Illinois University, Carbondale, Illinois. Dr. Remington has trained in Clinical Pastoral Education at Boston City Hospital, Worcester (Massachusetts) State Hospital, and Saint Louis State Hospital.

Dr. Remington began taking flying lessons in a Piper J-3 Cub at age 16 and has been flying since that time. During his 45 years of flying, he has accumulated nearly 1,300 hours. He holds Private and Commercial Licenses and a Flight Instructor's Rating in a fixed wing aircraft; he holds a Private Pilot's License in Sailplanes and is working towards a Commercial License and Flight Instructor's Rating in that aircraft. Jim won the Outstanding U.S. Male Pilot Award in his senior year in college; he held a Private Pilot's License in the Republic of the Philippines and flew an airplane as a missionary in that country. He taught flying and ground school courses at Parks College of Aerospace Education (Saint Louis University) for two years.

Dr. Remington serves (volunteers) as an officer and Mission Pilot/Chaplain in the Civil Air Patrol (auxiliary of the United States Air Force). He continues as a volunteer Interfaith Airport Chaplain at St. Louis Lambert International Airport, St. Louis, Missouri. In the six years he has served as Airport Chaplain, Dr. Remington has counseled numerous persons on numerous issues, many of which related to flying on commercial airliners or military aircraft.

Mr. Remington holds memberships in the following organizations:

The American Psychological Association
The Missouri Psychological Association
Anxiety Disorders Association of America
Association of Aviation Psychologists
Association of Christian Counselors
Aircraft Owners and Pilots Association (AOPA)
Experimental Aircraft Association (EAA)
Professional Airshow Performers Association
International Association of Civil Aviation Chaplains
Mental Health Association of St. Louis
Soaring Society of America

Dr. Remington developed the following program from personal and professional experiences in flying. His is not a "textbook program;" his is a practical program based on real experiences and needs. The user of this program will feel the sensitivity and caring of the developer.

Dr. Remington's genuine love for people and for flying, supported by his spiritual conviction, results in a program that will help you reduce your anxiety about flying.

He is deeply, personally dedicated and professionally committed, as you will read at the close of "Welcome to You" found in the next section:

> "Welcome to the experience of the reduction of your anxiety and to the thrill of flying—to flying relaxed and confident and feeling secure—to soaring like the eagles."

Dr. Remington is a licensed psychologist in private practice as a counselor and psychotherapist in St. Louis, Missouri. In addition to the material you are about to use, he and his wife, Leona, have other books, cassette tapes, and other helpful and interesting items either available for immediate order and purchase or currently being developed. The developers of this material are also available as speakers on a variety of topics. PLEASE TURN TO APPENDIX A FOR MORE INFORMATION. Dr. Remington may be contacted by calling (314) 533-4150.

3

TESTIMONIES

This program was extensively tested on people like yourself who were experiencing fears and severe anxiety about flying on a commercial jet airliner. They volunteered to test the program on themselves. They found from personal experience that the program really works!

These people were eager to share their success of overcoming their fears and anxieties about flying on commercial jet airliners.

Their testimonies--to their success and the effectiveness of this program--are made available to you.

My fear of flying began to develop suddenly after years of enjoying flying. The anxiety worsened with every trip, to the point I thought one day I would have a heart attack in flight.

Flying is necessary for my business and sometimes clients were with me. I stopped all pleasure flying for myself and avoided business flights whenever I could. The image of me on a plane and how out of control I was had to have made my clients wonder if I was capable of handling their account.

After completing this program, on my first trip to New York I experienced 1½ minutes of anxiety and <u>none</u> on the return flight!!

There's a whole world out there that I'm going to see.

Dear Dr. Remington .. Jim,

There has now been an eight - nine month period since I last wrote you. My last sentence turned out to be prophetic - I just returned from Europe! I logged eight flights and 23.5 hours in the air. This would have been totally unthinkable without your help. It all was entirely anxiety free. This was no doubt the most wonderful vacation of my life. I really owe you a lot for your help and continued support.

I'll keep you posted on my future travels, which are now a certainty.

Best success to you, as well.

Happy Flying! (Hope you don't mind that I stole that phrase.)

Ms. B. J.
Marketing Specialist

Dear Jim,

Thank you for helping me fly without fear.

The elation experienced during the flight is still with me. I have played my tape several times this week, and each time I relive the peace and exhilaration that I felt in the air.

I now have a "special place" to which to retreat whenever I am anxious--it is in the airplane looking out into the clouds.

Congratulations on your success and thanks for mine.

Mrs. C. K.
High School Teacher

Dear Dr. Remington

I just completed your ABC Program <u>*and*</u> *my very first enjoyable flight. I am thrilled to tell you that it (and I) was a fantastic success!*

I am a theology student and after reading the scriptural quotations you chose for your "Welcome" and the spiritual theme that flows through your book, I feel like you are a kindred soul. That feeling remained with me throughout the book and during my flight.

I followed your instructions in the "Welcome" portion of your program to first read the program in its entirety and then to re-read the program highlighting those areas that were relevant to me. When doing this second step, I found something surprising. Those areas that I highlighted were mainly regarding control. This new found realization has evolved into a spiritual revelation for me. My fear was not in giving control to the airplane, pilots, or flight crew - it was in giving control to God!

Your suggestion for using imaging and affirmations was extremely helpful. During my successful flight after completing your program, I envisioned and quietly affirmed: God is in control of this airplane and all its mechanisms and is working through the minds and hands of the pilots and the flight crew. God is in control of weather conditions. God is in complete control of me.

I believe that faith is the knowledge that God is in complete control of everything and everyone in His universe and there is nothing to fear. Thank you, Dr. Remington, for reminding me of this truth. Your book is more than just another book on the fear of flying - it is a lesson on faith in God and faith in oneself.

Bless you.

J. M.
Theology Student

Dear Jim and Leona:

I am a professional architect and have a substantial amount of success with my designing.

Unfortunately, I have been unable to design a way for me to fly to many parts of the world. My fear of flying was constricting my professional opportunities in a significantly severe way.

Your program worked for me. I now not only have no dread of getting on an airplane, I actually do a lot of thinking about my designs and perform some preliminary sketching while flying.

I continue to be amazed and grateful how my professional opportunities are expanding with my fear of flying decreasing.

Thanks to you and the program,

Mr. A. P.
Professional Architect

Jim

I really felt more in control of my thoughts, like you said in the program. I choose and control my thoughts. There were so many helpful ideas, my flight was successful, my anxiety was very low compared to how I felt on previous flights. I really enjoyed this one. I also felt more comfortable about airplanes.

M. J. H.
Meat Cutter

Dear Jim

I wanted to take a moment and tell you how wonderful it feels not to fear the thought of boarding an airplane.

My recent promotion has required many hours flying to from my destinations throughout eastern United States. I found myself making excuses, cancelling scheduled appointments and creating situations to avoid flying. I was to the point of taking a lesser position or even seeking employment elsewhere.

However, I had the good fortune of knowing an individual through work who told me about this exciting book titled "ABC How To Fly Without Fear." Without any hesitation, I decided to take the course. The course turned out to be the single mort important thing I've done in my life. It has given me two things I have always wanted, peace within myself through Jesus; and the ability to fly anywhere without any fear.

I appreciate the time we spent together and wanted to thank you for fulfilling a significant gap in my life.

Thanks again,
Mr. H. S.
Corporation Vice-President

I approached this program with an admitted professional skepticism. It seems that almost every program on the market these days is advertised as "built upon the foundation of proven educational and learning principles." As a professional adult educator, I know that most of these claims are simply not the case. A high disparity exists between what is said is done and what is really done. This incongruence between what is said and what is done becomes readily apparent as the student becomes involved in the material.

After I was involved in this program, my professional skepticism dissipated. Dr. Remington does in reality use proven and the most current educational principles and learning approaches throughout the total program.

I applaud this program from the perspective of utilizing the most effective and sound educational principles of adult education, and will be happy to have my evaluation and compliment put into print.

Dr. E. P.
Adult Educator

4

WELCOME

YOU ARE PERSONALLY WELCOMED TO THIS PROGRAM THAT WILL ASSIST YOU TO REDUCE YOUR ANXIETY ABOUT FLYING

This program is all about focus. The success of this experience depends almost totally on you and your focus. At this initial point, please pause . . . and make the following commitment: I COMMIT MY FOCUS TO THE NOW, AND TO FOLLOWING IN DETAIL THIS WHOLE PROGRAM. During your involvement with this program, do not allow yourself to "wander to the past" or to "jump to the future," except as the program specifically guides you in that regard. Your commitment is to work the program in its design of the now.

Let's not waste too much time on definitions. We could consume lots of time and effort defining concepts like "fear" and "phobias" and a huge bunch of related terms. Instead, we note—and respect—that for whatever reason or reasons, you have feelings that are negative to flying on a commercial jet airplane. Your feelings may fall somewhere within the wide range of a vague uneasiness to a discomfort to a worry to an intense dread and repulsion about flying. Let's simply call wherever you are in this wide range of feeling "avioanxiety," and get on with the more important challenge of reducing that anxiety.

A million people fly everyday in airplanes. Perhaps you label yourself "weird" because you are not enthused at this point about flying. Sorry, you are neither unique nor weird. It is estimated, probably conservatively, that 25 million people share some degree of what you are feeling about flying.

This program is about focus. And one important focus is on control. Accept—because it is true—that you possess the ability and capability to control your body, to control your attitude, and to control the essential you. Right at this moment, you are in control of yourself because you decided to actively involve yourself in this program that will reduce your anxiety about flying. You have everything you need to control yourself, to become relaxed, to become confident, and to feel secure.

You probably are starting out with a heightened sensitivity. You are particularly sensitive to sights and sounds and shakes and smells. This program starts with you where you are.

This program is divided into seven phases of the flight experience: from preparation for the flight at home to walking into the terminal building at your destination, smiling and happy. Anxiety reduction success will be experienced at the end of each phase and then at the successful completion of your flight. You will be systematically rewarding yourself throughout the program and the flight.

Four main anxiety reducing "tools" or techniques are used in the program: relaxation, deep breathing, meditation, and affirmations.

The two parts of RELAXATION are these: one is to select "in your mind's eye" a special focus of relaxation for you—your special place, person, experience, etc. In your imagination you will be going to that special focus during your flight. A second part of your relaxation is to have available <u>your</u> <u>kind</u> of relaxing music and a device on which to play that music.

DEEP BREATHING is a valuable assistance. It results in fantastic physiological and psychological experiences. First, <u>establish</u> <u>a</u> <u>rhythm</u> to your breathing. This is critical to the overall value and success of deep breathing. <u>Concentrate</u> on <u>inhaling</u> <u>air</u> . . <u>hold</u> . . <u>hold</u> (suggested slow count 1, 2, 3, 4, 5) . . <u>exhale</u>. Repeat. Repeat. The most effective breathing is with the lower lungs. Attempt to breathe at the lower lungs, then allowing air to fill your upper lungs.

Then you may wish to include words or very short phrases. Examples: "I inhale security; I exhale anxiety." "I breathe in relaxation; I breath out nervousness." "In . . Peace; out . . tension." "In . . Spirit comfort; out . . human aloneness."

MEDITATION has special meaning to persons who have a vital quality of the spiritual in their lives. Meditation is closing out the outer world of distractions, limitations, bad memories, negativity, and all other unhealthy stimuli; meditation is moving into the inner quiet or the calm within, often referred to as "the silence" or "secret place." Some people who meditate have no inner mental focus, they simply enter a "receptive void." The belief with this approach is that a true receptivity will result in our obtaining what we need the most at this time.

The development and repeating of AFFIRMATIONS are a critical ingredient in this program and in your flight. An affirmation develops from you reaching down into the depths of your being to your deepest core. You bring up from deep within you the most profound and practical truths about who you are and what you are doing. This is an opportunity to take advantage of <u>your spiritual strengths</u>. An affirmation is the development and the repeating of a truth about you and what you are doing, whether you believe all, some, or none of that truth at the time you begin the affirmation. In time, that total truth will become your total truth at your conscious level.

A repeating theme, as you continue your progress and growth in reducing your anxiety about flying is knowledge, observation, confirmation, and resultant trust.

The three primary focal points are RELAXATION, CONFIDENCE, and SECURITY.

There are throughout this program frequent changes in "tenses"--changes of "you," "I," "my." This is deliberately done so that program and your involvement become more and more personal to you--in truth, <u>more and more you</u>.

Holy Scriptures say, "I bless the Lord: O Lord my God, how great You are! . . . Whose clouds are chariots . . . Who rides on the wings of the wind." (Psalm 104:03)

And, " . . . they shall renew their strength . . . they shall mount up with wings like angels . . . and not be weary or faint." (Isaiah 40:31)

This whole program, then, incorporates and combines

the mind-body-spirit.

Welcome. Welcome to the experience of the reduction of your anxiety and to the introduction to the thrill of flying, to flying relaxed and confident and fully secure, to soaring like the eagles.

Jim Remington

5

INTRODUCTION FOCUS

SIX TOPICS WILL BE DISCUSSED IN THIS SECTION:

- THE VALUE TO YOU COMES FROM YOUR STUDY

- YES, YOU ARE IN CONTROL

- TWO SUBJECTS OF TECHNOLOGICAL ADVANCES

- QUESTION OF FLYING ANXIETY RETURNING

- MAKE YOUR SELECTION

- GROWTH BEYOND THIS PROGRAM

THE VALUE TO YOU COMES FROM YOUR STUDY

Few persons, if anyone, would challenge the statement that we have so much to do in such a short time! Of perceived necessity, we hurry, hurry, hurry, HURRY. This strong push to hurry often translates into a tendency (temptation?) to "quick read" material.

The material in the following pages is an educational book, but it is more than that. This book is more precisely, a study manual—a study manual of a program.

Persons who experienced the tremendous success of using this program urged us to pass on to all future program users this important point: start by reading all the material from cover to cover. Read each page with an intense interest.

Then begin reading as study. Study carefully each topic, subject, and phase of flight. Then, you are ready to "make your selection" as will be discussed in detail in the next few pages.

The point seriously stressed is this: you will do yourself a major disservice if you approach this material with anything less than a determination to study it thoroughly.

Incidentally, previous users of the program have told us that they have re-studied the manual some length of time after using the manual during a flight or flights, and to their surprise gained something new and valuable. We suggest that you do the same.

YES, <u>YOU</u> ARE IN CONTROL

There is a particular feeling frequently held in common by people who are anxious about flying: the fear of not being in control.

Whether you are a successful (and powerful) business person who has control over a large number of people, makes major decisions, and directs huge sums of money; a student who controls how seriously you study; or anywhere in between these two, once you sit in the seat of the airplane, a feeling of no longer being in control may sweep over you: "The airplane controls me. The pilots control me. The weather controls me. The (<u>whatever</u>) controls me. I no longer am in control!" This is a very unpleasant fear.

This may be a commonly held and unpleasant fear, but it simply need not be true. The control that really counts is the power you possess to give away the control. You give your control away to trust, a complete trust in the airplane, flight crew and cabin attendants, radio communication and radar coverage, and so on. Strangely, the more control you give away in trust, the more control you, in fact, retain. I trust in your control. You enjoy flight the most when you invest your confidence in the marvelous machine, the highly skilled and dedicated pilots and attendants, and the support personnel on the ground. Note that you are in total control of whether <u>you</u> invest your trust and entrust your confidence in everyone and everything having to do with your flight.

There is a second sense in which you remain very

much in control; you are in total control of the attitude you have about the flight. When your attitude is one of expecting a decrease of your anxiety about flying and an increase in your enjoyment of flying, the likelihood is raised to the maximum that, in fact, you will experience a decrease in your anxiety about flying and an increase in your enjoyment of flying.

Yes, you are in control!

TWO SUBJECTS OF TECHNOLOGICAL ADVANCES

You have placed confidence in me by purchasing this program and by trusting that use of the program will reduce your anxiety about flying and will increase your enjoyment of flying. Your trust will not be betrayed, so we honestly share two subjects that modern technology has <u>not</u> <u>yet</u> fully conquered. We do note that considerable progress has been made; the rapid and continuing advances of technology will move progress in these areas even further for the benefit of all of us who fly. One subject is "clear air turbulence." Clear air turbulence may cause the aircraft to bounce abruptly and sometimes to change altitude quickly.

Technology has not developed as far as is possible, so these invisible air currents are not fully detected by avionics, that is, a mechanical indicator to alert the pilots that the airplane is approaching clear air turbulence. Tremendous, exciting, and rewarding research is focussed on this indicator. At any time now, an advanced indicator will be available to point out that the airplane is approaching clear air turbulence so the pilots may simply guide the airplane to remain in the usually stable air.

The second subject is referred to as "wind shear." As the name states, wind shear is abrupt wind speed shifts or quick wind direction changes. Wind shear may cause the aircraft to make rapid changes in position.

Technological advances will produce an avionic, or mechanical indicator, that will call the pilots' attention well in advance of a wind shear situation and location. The indicator will combine the sophistication of three most-modern technologies: infrared sensors, Doppler radar, and laser-Doppler radar. At any point in time now, an advanced indicator will be available to inform pilots of wind shear, so the pilots will simply fly around that situation and location.

This advanced indicator is of particular value—two values in one, actually: (1) an alert to approaching clear air turbulence and (2) to inform the pilots of the existence in a specific location of wind shear.

It is important to note that either clear air turbulence or wind shear occurring during your flight is rare.

And, it is highly probable that before this manual is printed, those avionic indicators will be developed and installed and in use on your airplane during your flight.

An additional note is that constant and dedicated research is invested in our whole exciting field of aviation. Valuable developments will become realities and functioning in just the short period between the time this book is written and appears in print to be purchased. See how important you and your traveling by commercial jet airliner are.

QUESTION OF FLYING ANXIETY RETURNING

A relevant and logical question is whether one's anxiety about flying will return on the next flight. "I made the flight using the program, and exactly as you said would happen, my anxiety about flying decreased. But, will that anxiety return with my next flight?"

Almost certainly the answer is "Yes" . . . but it is "Yes . . . some." The key word is "some." Chances are your anxiety about flying developed over many years. It would be expected, then, that no anxiety about flying may be totally resolved and certainly no severe fear about flying may be completely cured by any program, however effective, and by any one person, however resolute and determined, during one flight.

There is a belief in this program that thoughts determine situations and events; we bring into reality that very thing on which we concentrate in our thinking. What is produced reflects the way we think. We should be cautious, then, not to bring something negative into reality simply because of the power of negative thought. The habit of negativity can sneak up on us and become a part of us, long before we realize what is happening.

> *Negative habits are not easily broken. Miracles happen, but to most of us miracles occur slowly. Expect your habit of negativity to be broken; anticipate the miracle of your increasing focus on the positive to occur, but be patient while all of this is taking place in you.*

It is acceptable to briefly consider that some anxiety may reappear on the next flight following one successful flight. This is easy to work with: re-study the program and re-apply all your selected helps on the second flight. Do this on every flight, as needed, until your anxiety about flying is reduced to your satisfaction or totally disappears.

MAKE YOUR SELECTION

We strongly suggest that you study the whole program well in advance of your flight. PERSONALIZE THE SECTIONS, TECHNIQUES, AND APPROACHES BEFORE YOUR FLIGHT—THESE MAY BE YOUR PRIMARY FOCUS DURING THE FLIGHT. The experience of people using this program reveal that most users have <u>specific concerns</u> about flying and that <u>particular phases</u> of the trip trigger this anxiety. Thus, most users of the program underline or highlight sections of this book that they feel will be of particular help. Be flexible, however, to use more than your selected sections once into your flight, if you feel the need.

A lot of help is here for you. Use as much as you need at this time. Again, the design and intention of this program and your dedication to use the program as designed and your commitment to change will reduce your anxiety about flying and increase your enjoyment of flying.

"How much time should I invest in this program before my flight?"

This is an excellent and very pertinent question. The answer involves the fact that each one of us reads at different speeds, our comprehension and retention levels vary, and we are not alike regarding the extent of our avioanxiety.

The thoroughness and seriousness of your study and use of the program as presented are highly correlated with the effectiveness of the program for you.

Read and study the whole program. Go over the entire program once more, highlighting the parts of particular application to you and your concerns and needs.

Persons testing the program indicated that the time they spent reading and studying the program before the flight ranged from 4 to 8 hours. They also stated the importance of making time to be very familiar with the program before the actual flight. Most persons said that it is preferable to start reading 10 days to 1 week before the flight.

GROWTH BEYOND THIS PROGRAM

People with anxieties about flying tested this program before we put the program on the market. We used statistical computations in evaluating the program's effectiveness in reducing anxiety about flying and increasing enjoyment of flying. The reduction of anxiety about flying was calculated, as was the enjoyment of flying.

These evaluating persons told us in writing about the program's results. As we carefully analyzed their written responses, we slowly saw a whole new result emerging, one that we had not even considered: as people reduced their anxiety about flying, they also became less anxious in other parts of their lives; as their enjoyment of flying increased, their enjoyment increased in other areas of their lives.

There exists the great potential for growth beyond this program in your life by successfully completing this program.

MAKE IT SO!

6

FINAL CHECKLIST BEFORE BEGINNING THE ACTUAL WORKING OF THE PROGRAM:

You are almost ready to begin working the program in detail, beginning with Phase One. The only remaining items before Phase One are the following:

1. Write down your external and your internal motivators.
2. Describe your "secret place."
3. List your relaxation music by name.

MOTIVATION. The higher your motivation to be successful, the higher the probability that you will be successful. An excellent guide in this regard is the "Forms of Transportation—A Comparison" found in Appendix B. Valuable factual and emotional insights and revelations often result from completing that document. You may turn to that section now and do it, or review if previously completed.

Please indicate below both your external and your internal motivations. For example, your external motivation might be "To arrive at a place where for personal or business reasons I desire to be," or "To be with my friend or friends." Examples of internal motivation could include, "To reduce my anxiety about flying," or "To boost my ego that I can make a significant change in my life."

Now, please, write your external and internal motivations.

MY EXTERNAL MOTIVATION(S):

MY INTERNAL MOTIVATION(S):

MY SECRET PLACE. Describe in detail the place in which, person with whom, or event at which you experience the most peace and relaxation; the exact situation in which you can be the most calm. This may be an actual situation from your past; this may be a fantasy of your present; this may be a projection of your intention of the future. WHAT IS IMPORTANT IS THAT YOU DESCRIBE IN DETAIL "THE RELAXATION SITUATION." YOU WILL BE "GOING THERE" AT POINTS IN THE PROGRAM.

MY SECRET PLACE:

MY RELAXATION MUSIC. Listening to your selection of relaxation music is integral to the program. List below the name.

MY RELAXATION MUSIC BY NAME:

YOU ARE NOW READY TO BEGIN WORKING THE PROGRAM AT PHASE ONE—GO FOR IT.

7

PHASES OF YOUR FLIGHT

PHASE ONE: PREPARATION AT HOME: LITTLE THINGS MEAN A LOT

1. TICKETS. Purchase tickets well in advance of your flight. When purchasing your tickets, insist on the seat location most comfortable to you. We strongly suggest that your seat be over the wing where, when and if you choose to look out the window, you will mainly see the wing.

You are encouraged to look at the wing while you are on the ground, and to look at the wing, when you feel comfortable to do so, while you are in flight. A window seat over the wing usually provides a view of the three important operating parts on the wing: the use and positions of the flaps; the use and positions of the ailerons; and the use and positions of the dive brakes. These will be discussed in detail in Phase II.

These three operating parts are more difficult to see from an aisle seat. However, some people prefer an aisle seat in the early stages of working this program. Persons who select aisle seats often do so because they do not have to look out the window directly to the ground and they do not like the feeling of being on the "downside" when the airplane banks, that is, the airplane turns in the direction of the side of the airplane they are sitting on and that side of the airplane is lower than the other side. Also, the aisle seat feels less confining to them, and this is a very big consideration to some passengers.

The window seat, again, allows a view of the wing and the operating parts on the wing. Persons who select window seats experience a sense of pride in being able to visually identify and watch the flaps, ailerons, and dive brakes work. There is an especial sense of pride in sharing your knowledge of the airplane and its flight with your traveling companion. You will know the "whats" and "whys" of the flaps and ailerons and dive brakes.

SEAT SELECTION OPTIONS. If a person is large and/or feels very uncomfortable about sitting in close proximity to other passengers, there are two options to consider: one is to pay the additional ticket cost and sit in the first class section. The seats are larger there; fewer passengers are in that location.

An addition to that option is to purchase two adjacent seats in the first class section. The arm rest usually either may be removed or tucks out of the way into the seats.

A SPECIAL NOTE. There is an increasing responsibility being laid on passengers who sit at and near the various exits on the aircraft. Persons who sit at or near all the exits make an unwritten commitment that they have the physical and mental capacities to follow the directions of the flight crew and cabin attendants, and, that they—the passengers—will retain their emotional composure to open the exit doors and to perform any other emergency procedures. At the time this program is developed, there is a clear moral imperative regarding the many exit doors and other emergency procedures. We anticipate that some legal responsibility in this regard will in the near future be applied. Please consider this when making specific seat location reservations.

2. DETAIL PLANNING. Plan in advance and in detail your projected travel arrangements, including (1) the packing of previously selected clothes, books, etc.; (2) the drive or ride to the airport; (3) checking in; and (4) arriving at your airline waiting area.

> PLAN TO ARRIVE AND BE SEATED IN THE WAITING AREA AT LEAST ONE HALF HOUR BEFORE BOARDING TIME—PROBABLY ONE FULL HOUR BEFORE THE TIME YOUR FLIGHT IS SCHEDULED TO DEPART.

3. LIST. Well before you start packing for this enjoyable flight, make a careful list of everything you need to have and do. It is a good idea to have your cassette player (batteries checked in advance) and relaxation music ready. Listening to relaxing music will be an integral part of this program and the success you will attain during this flight.

Air-to-ground telecommunication is becoming more and more attractive and, thus, more and more available. There is the ever increasing likelihood that your airplane will have a telephone that you may use to call anywhere while you are in the air. If you think that you might want to make a call while on your flight, call the airline you are flying with and obtain exact information as to what you will need to do in order to use the phone while you are working this program. (Phase Five discusses the use of the phone while in flight in more detail).

4. SPECIAL FOOD IN FLIGHT. If you need special food, especially for medical or religious purposes, be sure to let your airline know this several days before your flight.

5. COMPLETE PLANNING. Complete as much as possible all arrangements for transportation; places you will be staying; persons and sites you will be visiting; events you will be attending; time frames and timetables; and realistic estimate of costs. Plan now so there will be no "surprise" then.

You will want to be assured that your place, animals, plants, and so on, are in capable hands while you are gone. At least one person should know how to reach you during all the time you are away.

6. TRAVELING COMPANION. It is desirable to have a person make this important airplane trip with you. Ideally, this will be a person you trust and relate well to — maybe someone near and dear to you.

Especially important is that your selected traveling companion fully understands all aspects of this program, is sensitive to your concerns, is uncritical of your anxieties or fears, and is supportive of you and your following this program in detail all during this flight in which you will reduce your anxiety about flying.

7. AIRPORT VISIT. A neat idea is for you to take a trip or two to the airport. Watch the gracefulness and awesomeness and safety of airplanes taking off and landing. Tell yourself how fortunate you are that you soon will be on an airplane and that you will be taking off and landing and having an enjoyable experience.

There is frequently a negative feeling, or outright fear, about going to the airport. This is a fairly common experience with people who have an anxiety about flying. If you have this negative feeling at all, please do the following: set up a precise schedule of slowly and systematically arriving closer and closer to your airline waiting area. This is technically called "de-sensitization." Its importance is great. Set up a precise schedule. You may find that you are able to move or complete several parts of your de-sensitization schedule all at once. That certainty is acceptable, but insist of yourself that each part of the schedule is completed successfully as you planned that schedule.

Here is a suggested de-sensitization schedule:

1. Simply drive past the airport.
2. Go to the airport and observe the airplanes taking off and landing from one or more locations near or on the actual airport.
3. Go to the airport, park, and just enter the terminal building.
4. Go to the airport, park, enter the terminal building, and go as close to your airline's waiting area as is possible.

You may need to use deep breathing and affirmations and imaging in this schedule.

If, of course, this is not a problem for you, skip this part and proceed to the next part in your pre-flight, at-home, preparation.

Go to the waiting area of your airline. Sit down and

become familiar with the surroundings (noises, a seat where you can look out a window at your airplane, people moving, door to the jetway, counter, etc.) and reflect: imagine yourself in this waiting area, waiting for your flight to be called; and imagine the specific activity you will be involved in at that time.

Shift your attention for just a moment to the possible smell of kerosene. You may sense this odor as you drive to the airport and while you are in the waiting area. This same smell may come to you as you are walking the jetway to the airplane, and maybe while you are sitting on the airplane as the airplane is being prepared to leave the gate. Kerosene is the fuel used in the jet engines of your airplane (like gasoline or diesel in your car or truck). It is the "stuff" that causes the jet engine to produce so much power.

Kerosene is a strong smell in and of itself. It is the predominant smell around an airport especially when the doors are open, as is the case when passengers are entering or exiting the airplane or when baggage is being loaded.

A nice approach, once you smell the kerosene, is to make a quick affirmation: "I bless that smell because the fluid of that smell will provide the fantastic amount of power in the engines to speed me to my destination."

IMAGE, that is, strongly imagining yourself following this program; image yourself becoming relaxed and even excited (in a positive way) about the flight.

You clearly image yourself, when your flight is announced, walking to the door, walking the jetway to the airplane, entering the airplane, finding your seat, storing any items you have with you in the overhead compartment or under the seat in front of you, and sitting in your seat.

Include in this full imaging of your actions a focus on what else you will be experiencing—knowledge, relaxation, observation and confirmation, and above all, the feeling of security.

8. CLOTHES YOU WILL BE WEARING. Select and wear casual clothes that are loose fitting, especially around your neck and waist; and wear comfortable shoes. Remember your sunglasses because even if the day is cloudy on the ground, you probably will go above the clouds into a bright, clear sky with brilliant sunshine.

9. HEAR, SEE, AND SPEAK NO EVIL. Don't allow yourself to hear or see any news, be engaged in any conversation, or think anything negative about yourself and the approaching thrilling trip on the airplane.

10. RELAXATION AND BREATHING EXERCISES. Begin practicing relaxation at home. Determine your "special place" of relaxation. When or how do you relax the most? You will be "going to that place" frequently during your airplane trip. So practice going there now. Think, feel, hear, smell, and totally experience your special place of relaxation.

Select a word, words, or short phrases that help you relax and repeat this relaxing word or words while deep breathing. Practice deep breathing at home, incorporating what you have found helpful.

Begin <u>practicing</u> <u>deep</u> <u>breathing</u> at home, as has been said. Most people benefit the most from deep breathing by sitting in an upright position with feet flat on the floor and hands laying in lap; the hands may or may not be touching. If possible, closing the eyes usually helps to "focus within." Inhale a breath of air . . . hold it . . . slowly exhale. Repeat . . . repeat . . . repeat.

As you have studied this manual, it is preferable that you have—at home—chosen appropriate <u>affirmations</u>. We strongly suggest that you write these affirmations in your manual while you are at home. Feel free to modify them or to make new ones, once you are in the air and are working each actual program phase of your rewarding trip.

11. NO ALCOHOL OR DRUGS. I determine that I will not take any kind of an alcoholic drink (or use any other non-prescription drug) at any time during my airplane trip. If you are taking a prescription medication, consult your physician about your medication and your flight.

12. RESTROOM? Many people find using the restroom on an airplane a less than desirable experience for a variety of reasons. You may wish to be cautious about how much you eat and, especially, how much you drink before the trip. Of course, you will plan on using the restroom in the waiting area before you board the airplane.

A few more words about the lavatory on airplanes. Passengers have shared five main concerns with using the restroom while in flight (you cannot use it while on the ground, of course).

1. The compartment is quite small. Because of the

confined space, items easily noted in larger restrooms are often difficult to locate and to identify quickly and properly. Everything found in larger restrooms is there; the smaller space necessitates different handles, labels, latches, doors, drawers, and creates different sounds and sensations.

2. The inside door latch is confusing to some people. In the best interest of you and the person approaching the lavatory you are in, you probably will want to fasten the door closed as soon as you are inside. When you latch the door from the inside, appearing in the little window on the outside is the word "Occupied."

3. Because this important "office" is usually located near the aft (rear section-no pun intended) of the aircraft, the noise level and vibrations from the engines are often substantially greater. Expect an increase in noise and vibrations when in the lavatory.

4. The combination of the enclosed area and the location in the airplane often produces a sensation of "swaying" motion. Certain human anatomy features, especially of the human male, require certain safeguards and actions not required in using a toilet that is not swaying.

5. Note this one point especially: the noisy sound of rushing air when the toilet is flushed is at least disquieting, if not outright frightening, the first time you flush the toilet in this unfamiliar surrounding. The sound of rushing air is simply a characteristic of the way the toilet-flushing mechanism functions. Nothing is wrong. One person reported, however, that when she flushed the toilet, the sound and rush of air caused her to think that she somehow

had mistakenly pulled the "escape" handle.

Lots of people use the lavatory on airplanes. What we don't know is what was their first experience in doing so. It is suggested that before the need to use the lavatory is too urgent, go into the lavatory and allow yourself time to become familiar with all the necessary items and how they operate. Make time to turn the unfamiliar into the familiar. And, as always, feel free to ask a flight attendant for any help, however trivial it may seem to you. The once unfamiliar lavatory will become familiar to you.

And now to the inner growth work—

SHORT FOCUS IMAGING/AFFIRMATIONS.

Again, start right now, at home, before your airplane flight, imaging yourself. Image yourself leaving your home, driving/riding to the airport, checking in, arriving and sitting in the waiting area relaxed and confident. You image yourself feeling secure that all the details of your exciting flight have been completed. The three main feelings you know that you will experience are RELAXATION, CONFIDENCE, AND SECURITY.

Develop your own AFFIRMATION, and start repeating it. Your affirmation might be like this one: I SHALL ARRIVE AT THE WAITING AREA FEELING RELAXED AND CONFIDENT AND SECURE.

YOUR AFFIRMATION: ____________________

____

____

LONG FOCUS IMAGING/AFFIRMATION: Start right now, at home, before your airplane flight, imaging yourself. Image yourself leaving the airplane, walking the jetway, and entering the terminal building. Your main image is your attaining a successful flight. Maybe you want to include the thrill of greeting or being greeted by a person or persons waiting for you to arrive, and/or the excitement of actually arriving and being at your destination. Again, the focus of your image is of you completing the successful flight.

Develop your own AFFIRMATION and start repeating it. Your affirmation might be like this one: I ARRIVE AT MY DESTINATION AFTER A SUCCESSFUL FLIGHT.

YOUR AFFIRMATION:____________________

____

____

All these little things, your preparations at home, mean a lot.

PHASE TWO: PREPARATION AT THE AIRPORT WAITING AREA: MAKING A FRIEND OF THE AIRPLANE

ARRIVAL. I have now arrived at the airport. I come to my airline waiting area and find the location of the door (Gate #) through which I will walk onto the airplane.

I find a place to sit where I can look out the window and see "my airplane" and where I am least disturbed by other people.

TRANSITION FROM PHASE ONE AND PHASE TWO

1. I want first of all to acknowledge successfully completing all of my preparations at home. My "test" of the success so far has <u>not</u> produced absolute feelings of relaxation, confidence, and security. Rather, I recognize that I am more relaxed, that I feel more confident, and that I feel more secure <u>in comparison with what I felt before</u>. My success so far, then, is that I realize that I am making progress. I ACKNOWLEDGE THIS PROGRESS, OR GROWTH, TO MYSELF (and maybe to my traveling companion).

2. I settle in my seat here at the waiting area. I do what I need to do to feel comfortable. I spend whatever time necessary to become familiar with my surroundings: I confirm that I am at the correct airline and at the correct gate through which I will pass to that "big, beautiful bird" out there. I confirm that the airplane I see out the window will be "my special friend" as we travel together. OK, I feel good about this place.

Realism dictates that a possibility needs to be addressed. At any point from arrival at the waiting area to actually being on board the airplane, a delay in departure may be announced. In rare instances, passengers may have to transfer to a different airplane. Although delays or plane changes are frustrating and upsetting, any delay is made in order to ensure that flight conditions will be at their best.

When a delay occurs, you are urged to practice your relaxation with an undergirding theme that you appreciate that everything is being done to benefit you and your flight. Please do not express your dissatisfaction to either the cabin attendant or the flight crew; most orders of delay come from other sources.

3. So for a moment now, I sit back, close my eyes, and relax. . . . Now, I'm ready to continue my progress in reducing my anxiety about flying.

MY ATTENTION TURNS NOW TO MAKING A FRIEND OF THE AIRPLANE. I open my eyes and I look out the window at the airplane that will be speeding me to my destination and on which I will be feeling secure.

I pause for this or a similar AFFIRMATION: WE SHALL BE GOOD FRIENDS.

YOUR AFFIRMATION:____________________

____

____

I attend, now, to the physical details of this airplane and of its flight:

1. There are four physical laws, four fundamental forces, four aerodynamic forces, that operate on this airplane.

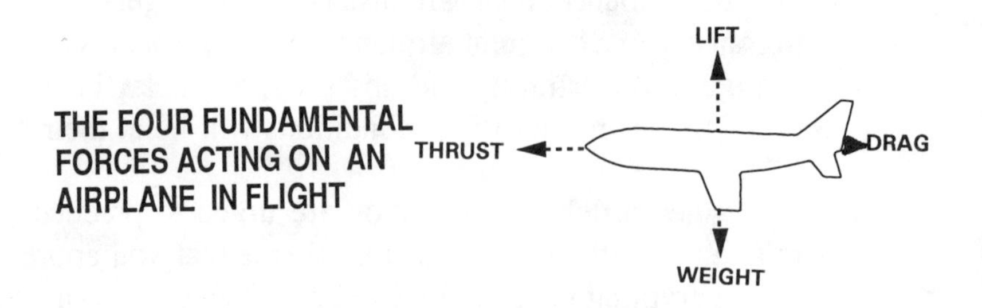

The physical laws are that the airplane flies when the lift is greater than the weight and the thrust is greater than the drag.

2. Lift comes from physical laws operating on the wing. It is difficult to see it, but if I could look directly at the wing from the end of the wing, I would see a configuration something like this:

A VIEW OF AN AIRPLANE WING, LOOKING AT THE WING FROM THE WING'S TIP

The thrust of the airplane through the air causes the air to flow over and under the wing surfaces.

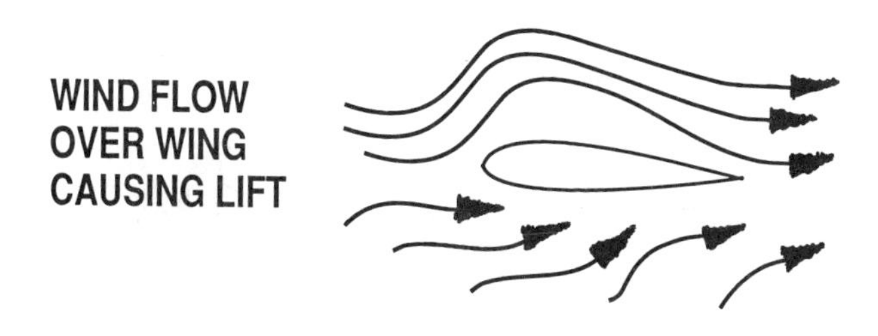

The air flows over the top surface of the wing causing a sort of vacuum or low pressure space. Thus the wing tends to raise: lift is in operation. The air flowing under the bottom surface of the wing "pushes" up on the wing. Another contribution to lift.

There are two simple and informative ways of illustrating lift: one is to extend your arm out the window of a speeding moving vehicle. (Passenger! Right side!) Note that when your hand is parallel to the wind flow, that is, exactly level and with the horizon or street and ground level, your hand and arm remain in that position with little or no effort on your part to hold them there. Slightly tip your hand up, and the wind flow will lift your hand and arm. That correctly illustrates lift.

A second illustration of lift is to hold a sheet of paper by both hands up to your face, flat, and just below the level of your mouth. Blow over the sheet of paper.

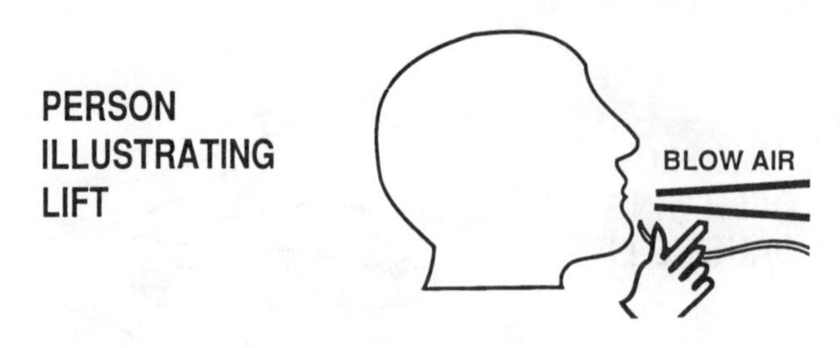

The loose end of the paper will lift. This also correctly illustrates lift.

Now, back to the airplane.

I may see that the wing appears to be thin and really not all that big. I wonder how the physical law of lift can occur? The answer is that because the airplane is going through the air so fast that the wind is correspondingly flowing over the wing rapidly. Thus, the wing does not need to be very thick or very big.

Someone, of course, raises the question, "But what happens when the airplane is taking off and landing and the air is flowing more slowly over and under the wing?" There are mechanical devices built into the wing that are used when the airplane is flying slower. Literally, the wing is "extended" or "enlarged" in those instances, so the same lift effect exists as the shape of the wing is changed.

3. I may be able to see what appear to be "little wings" on the trailing edge of the big wing, on the little wing at the back of the airplane, and on the tail sticking up. There are at least three "little wings" at those three locations.

LITTLE WING NAME & LOCATION	WHAT IT ALLOWS THE AIRPLANE TO DO	WHAT THAT AXIS IS CALLED
Ailerons on the big wing	Roll to the left and right	Longitudinal
Elevators on the little wing at the tail of the airplane (horizontal stabilizer)	Go up and down	Pitch
Rudder on the tail sticking up (vertical stabilizer)	Turn to the right and left	Yaw

The pilots in the cockpit have controls that move each of these three little wings. The airplane flies smoothly and makes perfect turns when all three controls are used at the same time and are coordinated in their use.

4. I may be able to see one or more of the engines. These are jet engines. The jet engine is relatively simple in design and in operation. The whole aviation field made sudden and dramatic progress and growth when jet engines became available. The jet engine is also a very dependable piece of machinery.

A jet engine looks and functions like this:

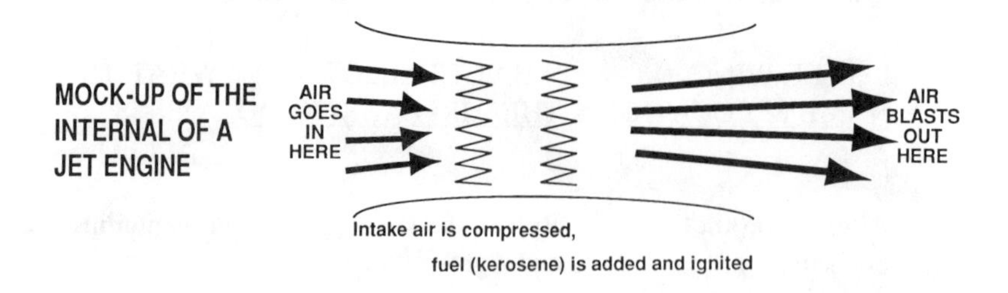

The blast of air at the rear of the jet engine is so much more powerful than the air coming in that tremendous thrust occurs for the airplane.

A jet engine uses a fuel that smells like and is kerosene. Sometimes that smell exists when you are around an airport and airplanes.

All those jet engines you see out there on the airplane are simple, dependable, and powerful.

Let's consider a definition of "friend." A friend is someone, or something, you know well, you accept each other as you are, and you trust each other and depend on each other. Your new friend out there—your airplane—is a carefully designed, constantly checked, mechanical marvel. That mechanical wonder out there has no doubt flown millions of miles; thousands of people just like you have experienced the thrill of flight, and so will you. And, that very special friend welcomes you aboard.

And now to the inner growth work—

SHORT FOCUS IMAGING/AFFIRMATION

I image myself, spending the remainder of my time in this place as relaxed, confident, and secure. I image myself, when my flight is announced, walking to the door, passing through the door, walking the jetway, entering the airplane, finding my seat, securing my personal items in the overhead compartment or under the seat in front of me, and sitting in my assigned seat.

I develop my own AFFIRMATION, and I repeat it. My affirmation might be like this one: THIS AIRPLANE AND I ARE FRIENDS. I TOTALLY TRUST THE PHYSICAL LAWS THAT CONTROL THIS AIRPLANE AND MY FLIGHT.

MY AFFIRMATION: ______________________

____

____

LONG FOCUS IMAGING/AFFIRMATION. I image myself completing this flight successfully. I image myself leaving the airplane, walking the jetway, and entering the terminal building.

I repeat my affirmation (maybe like this one): I ARRIVE AT MY DESTINATION AFTER A SUCCESSFUL FLIGHT.

MY AFFIRMATION: ______________________

____

____

OPTION: As I am now prepared to board the airplane, I have a couple of related options: I may call my clergyperson and have a prayer over the phone, or, I may request that the airport chaplain be called and come and have a prayer with me.

This is Jim sharing again personally with you.

I wish to share with you my especial desire right now to be supportive of you, and for you to know that I am supporting you all the time and in each and every phase of this total great program.

You are to be congratulated for all the time and effort, concentration and study-focus you have already devoted to the details of this growth program. Nice going! You are fulfilling the key role in reducing your anxiety about flying and in increasing your enjoyment of flying. My personal role, the contribution of my co-developer, and the design and dedication of this program are to assist you in attaining success. That is happening right now.

My imagining you successfully completing each phase on a short-focus basis, and you successfully completing the whole program on a long-focus basis, continues to be vivid, and I am totally confident that your success-focused-commitment will make these powerful images a reality in your life and your flying in confidence is increasing, even as you are reading my words.

My earliest affirmation made at the very beginning of this program persists to convey my support, and to bring other psychological and spiritual supports to you.

My affirmation again: "I AFFIRM YOU MAKING THIS AIRPLANE FLIGHT WITH A REDUCED ANXIETY ABOUT FLYING AND WITH AN INCREASED ENJOYMENT OF FLYING."

My deeply held personal spiritual belief system is that we are always within a Divine Plan—living within a Divine Blessing. With commitment and receptivity on our part, we are consistently and assuredly at the right place, doing the right thing, for our overall best good.

At this precise moment, then, you are at the right place, doing the right thing, for your overall best good. The end result shall be a decrease in your anxiety about flying and an increase in your enjoyment of flying-you are definitely moving toward flying in a commercial jet airliner, relaxed and confident and feeling secure.

You are on your way to soaring like the eagles!

Affirmations, imagings, and blessings, follow you and support you all the way.

MY FINAL ACTION HERE IN THE WAITING AREA: I practice my relaxation. I listen to my relaxation music. I focus on my feelings of relaxation, of confidence, and of security.

I AM RELAXED, CONFIDENT, AND SECURE WITH MY FRIEND AND ABOUT MY FLIGHT.

PHASE THREE: BOARDING THE AIRPLANE: GIVING YOUR PERSONAL BLESSING

"JIM WELCOMES YOU"

PRE-BOARDING DECISION. When the first announcement is made about boarding the airplane, the announcer will say that people with special needs may enter before the passengers are admitted in groups.

There is usually quite a mass jam of humanity when the passengers are admitted. Some people experience this physical closeness and the hustle and bustle as very unnerving.

You have the right to ask to board the airplane before the passengers are admitted in a group. You probably will not even be questioned about why you wish to do so. If you do happen to be asked, say something like, "I'm working on a fear of flying program, and the program requires me to enter the airplane this way." Nothing else will be said.

You also have the right for someone to accompany you while boarding the aircraft and settling in your seat. This probably will be the airport chaplain, your clergyperson, airline representative, or a special friend. Be sure and clear this in advance with the airline personnel at your gate. Again, simply tell them that the person accompanying you onto the airplane is a part of a program you are working to reduce your anxiety about flying. Nothing else will likely need to be said.

TRANSITION FROM PHASE TWO TO THREE

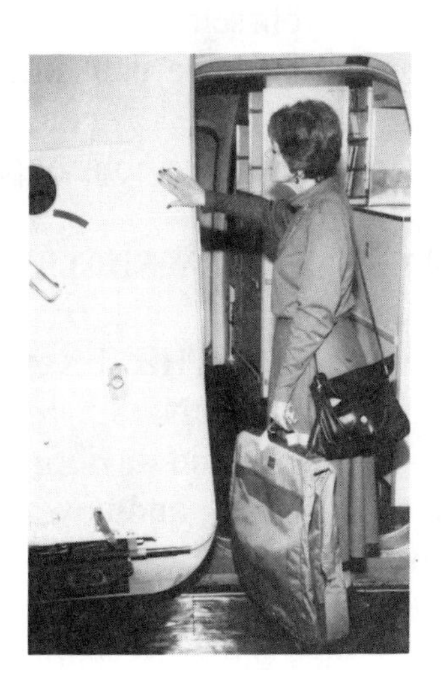

"PASSENGER BLESSING THE AIRPLANE"

1. You are now in the process of entering the airplane. We suggest that as you reach the door of the airplane, and you step into the plane, that you pause <u>very briefly</u>. Actually touch the airplane, and silently say something like this: I BLESS THIS AIRPLANE, THE FLIGHT CREW AND CABIN ATTENDANTS, AND OUR FLIGHT.

YOUR BLESSING: ______________________________

____

2. Find your seat, store your personal items, and be seated. You may place your items in the overhead compartment or under your seat in front of you. (Note: all items <u>must</u> actually fit under the seat.) MANY PEOPLE FEEL SECURE IN FASTENING THEIR SEAT BELTS JUST AS SOON AS THEY ARE SEATED.

3. Acknowledge to yourself (and your travelling companion) the success you have had during the waiting time and in boarding the airplane. Again, the recognition is based on improvement over previous experiences.

I FEEL GOOD ABOUT MY SUCCESSES.

NOW, WELCOME TO THE EXPOSURE OF A WORLD OF SIGHTS, SOUNDS, SHAKES, AND SMELLS. Expect —and welcome—a definite change from everything done up to this point; and now appreciate being on board the aircraft.

1. A good idea is to open the little ventilation valve over your head. This will bring fresh air down to you; and you may direct both the flow and the amount of air as you turn and change the valve. Many people feel better with fresh air on them. The air conditioning will come on when the

engine starts.

2. Take the diagram of the airplane from the seat pocket in front of you. Locate exactly where you are sitting, and the closest exits. Study this until you are comfortable that you know how to exit the airplane if a need would arise.

3. ALLOW THE SIGHTS, SOUNDS, SHAKES, AND SMELLS TO COME TO YOU. You will see an increase in number and speed of people, both inside and outside the airplane. You may see impatience and possibly rudeness.

You will hear sounds that might be unnerving to you until you are able to identify their source. There will be lots of "slamming" sounds. These slamming sounds come from the baggage doors under you being closed securely; doors and drawers in the kitchen being closed securely; the overhead doors (lots of them) being closed securely; and then the doors of the airplane being closed securely. Those slamming sounds equate to security for you.

You will probably feel the airplane shaking. Remember, the airplane is mounted on tires, just like your car or truck. The movement of a mass of humanity, food, and fuel being loaded, baggage added, and vehicles touching or pulling away from the airplane all combine to cause the airplane to shake. Those shakes equate to security for you.

Smells of food, engine exhausts, perfumes, etc. may reach you.

PAUSE TO FOCUS ON THESE SIGHTS, SOUNDS, SHAKES, AND SMELLS. NOW YOU CAN IDENTIFY

MOST IF NOT ALL OF THEM. ALL COMBINE FOR YOUR SECURITY DURING THE FLIGHT.

4. THINK ABOUT THE FLIGHT CREW AND CABIN ATTENDANTS. Every airplane will have a flight crew of at least two, sometimes three or more, depending on the size and complexity of the airplane.

If you arrived at the waiting area very early, it is quite possible that no pilots were in the airplane. They are not in the airplane because they are very busy. Before the pilots enter the airplane, they devote an intense study of <u>everything having to do with your flight</u>: weather, fuel, distance, weight of the airplane loaded, airports of destination, and a look at the airplane from outside the airplane. This careful visual check of the airplane is called "pre-flight."

A detailed FLIGHT PLAN is submitted by the captain—a plan with specifics concerning <u>every aspect</u> of the proposed flight. The governing facility over the final approval of the Flight Plan, and the continual monitoring of the Flight Plan while the airplane is in flight, is the FAA's <u>ATC SYSTEM</u>, the Air Traffic Control System.

When the pilots are in the airplane, they are busy checking and rechecking every gauge, radio, chart, reports of all kinds, and all the aircraft systems. THE PILOTS FOLLOW DETAILED CHECKLISTS. Nothing is left to chance or guess.

The cabin attendants are thoroughly trained, also. Feel totally free to ask anything of them; they are there to help you have the best flight possible. PAUSE TO FOCUS ON

THE TRAINING, EXPERIENCE, COMPETENCE, AND DEDICATION OF THE FLIGHT CREW AND CABIN ATTENDANTS.

5. BACK-UP SYSTEMS. Unlike cars, trucks, buses, boats, and motorcycles, airplanes have at least one back-up system for each individual system. This means that if the proper functioning of any one system—radio, hydraulic, electrical, oxygen—is called into question or if it malfunctions, the pilot simply changes to a back-up system. What a secure knowledge.

6. ENGINES. Each engine on your airplane develops more power than could hardly ever be needed; there are more engines on your airplane than needed. Look at the engines and feel very secure. You are secure.

7. RADIO AND RADAR COVERAGE. Every part of your flight—from getting ready to leave the gate of departure to arriving at the gate of destination—is closely monitored by radio contact and radar coverage. The pilots talk frequently, and radar is watching your flight constantly. Permission must be received prior to any change of the airplane, with the exception of an emergency. What a secure feeling this is. Information vital to your flight and other airplanes comes to your airplane and goes out from your airplane.

PREMOVING ACTIVITY. There may be a few minutes between the time you complete this program to this point and the airplane beginning to move.

1. Watch the cabin attendant as she/he demonstrates the seat belt, emergency oxygen, and exiting. Be comfortable

that you thoroughly understand everything said and demonstrated. The cabin attendants' talk and demonstration are required by the FAA (Federal Aviation Administration). This exact talk and demonstration is normal and routine.

2. Sometimes a passenger near us wishes to engage us in conversation. That attempt to make conversation is well intentioned, but probably not helpful to you at this point. Courteously and firmly inform the person that you are concentrating on a program and, with no disrespect meant, you will not converse with him or her. Remember earlier in the program, "A Very Basic Right?"

3. Practice your relaxation (music) and deep breathing.

And now to your inner growth work—

SHORT FOCUS IMAGING/AFFIRMATION

I image myself in this airplane as it taxies out to the runway and as it takes off. I image myself relaxed, confident, and, especially, feeling very secure.

I develop and repeat my own AFFIRMATION or use one like this: I AM COMFORTABLE IN WHAT I KNOW ABOUT THIS AIRPLANE AND MY FLIGHT. I AM SECURE!

MYAFFIRMATION: ______________________

____

____

LONG FOCUS IMAGING/AFFIRMATION

I image myself completing this flight successfully. I image myself leaving the airplane, walking the jetway, and entering the terminal building.

I develop my own AFFIRMATION or use one like this: I ARRIVE AT MY DESTINATION AFTER A SUCCESSFUL FLIGHT.

MYAFFIRMATION:______________________

____

____

PHASE FOUR: TAXIING AND TAKE-OFF: KNOWLEDGE AND TRUST; OBSERVATION AND CONFIRMATION

All opening parts of the airplane are securely fastened, the cabin attendants sit down (disappear from your sight), and the airplane is ready to back or be backed away from the gate.

TRANSITION FROM PHASE THREE TO PHASE FOUR

You congratulate yourself on the anxiety reduction all the way up to this point. I ACKNOWLEDGE THIS PROGRESS AND GROWTH IN ME.

TAXIING

1. As the airplane becomes ready to move, the lights usually blink and a new sound is often heard. This merely indicates that the electrical supply has been switched from an external source to the source within the airplane, and that the engines are now running.

You may hear, and possibly feel, a rush of air out of the overhead adjustable ventilators, especially if one or more of them is open, as will probably be the case. This will be only a momentary rush of air. The reason is that the air conditioning is now coming from the airplane, rather than from a piece of equipment on the ground near the airplane. That transition of air conditioning source often causes a momentary rush of air.

2. Sometimes a "strange sensation" is felt as the airplane backs away from the gate or is backed away from the gate

by a tractor. This is like the sensation felt when you sit in your car at the car wash. The brushes move past you. You have the sensation of moving, which of course you are not. This strange sensation will pass quickly.

3. As the airplane is moving ("taxiing") to the active runway (the runway from which you will be taking off), the pilots are following precisely a checklist of everything having to do with the functioning of the airplane and your flight.

If you look out at the wing, you may notice several things happening: the "little wings" out on the trailing edge of the wing (remember? ailerons) will move up and down; some flaps toward the middle of the wing (dive brakes or "spoilers") will come up and go back down. Especially notice that a part of the trailing edge and a part of the leading edge of the wing will be extended. These are flaps.

The ailerons will move often while the airplane is in flight; the dive brakes or "spoilers" will be used to slow the airplane down when it begins its early stage of letting down, that is, descending, and when the airplane has just landed. The flaps will always be used when taking off and when landing. Remember, they primarily help the wing to generate lift at slow speed.

On at least one airplane, the Super 80 or "stretched out DC-9", you may notice, as you look out your window, that below you on the wing are several triangular patches (often yellow or red in color) with what appears to be tape holding a short piece of yarn (a small cord, usually red).

You will rarely see this, but when you do you will

know that airplanes are sprayed with a de-icer fluid whenever any possibility of ice exists. The mechanic, when spraying the airplane with de-icer fluid, uses that device, in part, to establish that the de-icer fluid has been properly applied. That "thing" on the wing has absolutely nothing to do with the flying capability of your "big, beautiful bird"—ignore it.

4. In many airlines, the cabin lights are dimmed while the aircraft is taxiing out and often remain dimmed until the airplane is in level flight. You probably will not notice it during the day. You will notice the lights if they are dimmed during night flight.

You have the option of turning on, or leaving on, your overhead light during this taxiing phase, as well as during the entire exciting flight.

5. Sometimes the taxiing speed will vary. At times you may stop and then resume moving. What you probably do not see are airplanes in front of you. The speed of taxiing is mainly determined by the distance to other airplanes ahead of you.

6. Usually, the runway your airplane uses for take-off is the runway facing into the wind. A bit of trivia: at the landing end of every runway, there are two large numbers. The third number, not appearing, is understood to be a zero. These two numbers indicate a magnetic heading of that runway. For example, a "30" means that that runway points northwest, or 300 degrees. An "L" or "R" following the numbers indicates that there are parallel runways and that you are on either the left "L" or the right "R" runway. Now, doesn't this bit of trivia really excite

you?

7. Again, depending on airplane traffic in front of you on the ground and airplanes in the air approaching the airport to land, you may stop at the edge of the active runway or you may move from taxiing right onto the active runway for immediate take-off.

8. A pilot will announce to the cabin attendants to prepare for the take-off. This announcement simply reminds the attendants to sit down and fasten their seat belts, exactly as you are doing. This announcement is required by the FAA and is normal and routine.

THE TAKE-OFF

1. Expect a change in speed, a change in sound, a change in the attitude of the airplane, and a change in the pressure you are feeling in your seat (the airplane's and your's), as the airplane takes off. "Attitude" means the position of the airplane.

You will experience a definite increase in the sound of the engines, vibrations in the airplane, and an acceleration of speed. You will clearly feel yourself "pushed" back in your seat. Many people find this a particularly thrilling part of their airplane flight. I THRILL TO THE POWER OF THIS AIRPLANE.

As the speed increases so the lift increases, and the airplane becomes increasingly closer to flying. The first change in attitude you may notice is the front part of the airplane coming up. That indeed is what is happening. This attitude is called "rotation." Rotation means that the airplane is tipped back in lift-off configuration: it is literally rolling on its main gear with the nose gear no longer touching the runway. The airplane will be held in this rotation attitude until it literally flies itself off the ground. You are now flying. What a thrill!

You may sense on lift-off and climb that the attitude is very steep. That really is not the case. Most take-offs and climbs are around 20 degrees. The airplane is climbing at that attitude in part to protect the people on the ground from the noise. The aviation community is sensitive and really cares about the noise factor to people who live in communities near airports. Remember, also, that only part of the engine power is being used. A lot more engine power is available, if needed.

Very quickly after take-off, you will hear a "churning" or "cranking" sound followed by a "clank" or "thud" sound. The first "clunk" or first several "clunks" you may hear very soon after the airplane is climbing is that of the landing gears (wheels) retracting into the airplane (first clunk) quickly followed by one or more "clunks" as the doors fold over the landing gear. The wheels are securely positioned inside the body (fuselage) of your airplane and the doors are closed, thus, snugly protecting the wheels until they will be lowered in the landing approach. The airplane is now partly "clean."

At different times during the climb, you may hear other sounds and perhaps feel the attitude of the airplane change a little bit. There is often a series of "churning" sounds. The flaps are being retracted. Look out the window, if you wish, and visually determine whether the flaps are partially or fully retracted. Generally, the flaps are retracted in a series of steps (degrees). When the flaps are fully retracted, the airplane is now "clean."

The airplane will continue to climb until the assigned cruise altitude is reached.

Note that landing gear flaps sounds vary from airplane model to model. In some models, you can quickly and easily identify the landing gear retracting, and extending; in other models, it may be difficult to do.

You may look out the window at the leading edge and the trailing edge of the wing. No longer do you see exposed parts inside the wing nor will it be possible to look through the opening between the wing itself and the extended flaps and see the ground—the flaps are completely retracted—the wing now looks totally intact.

As the airplane is climbing in altitude, the change in pressure may cause some problems with your ears. This is a physiological response to pressure changes. Chew gum, swallow, yawn enthusiastically, hold your nostrils and "push" air. This, too, shall pass. You may wish to refer to the brochure listed in Phase Six.

The airplane will make many turns before it levels off. The pilots are following directions as to exactly which way and precisely at what altitude to fly. Remember, you are being followed by radar coverage. Radar coverage keeps a distance between you and all other airplanes. Sometimes your pilot will want to fly around a storm or severe weather. They will ask, receive permission, and change the direction and/or altitude of the airplane.

A review, when the airplane is taking off—landing gear down and flaps extended—you are considered in aviation talk as "dirty." When you level off—gear up and flaps retracted—you are considered "clean."

You are now flying along very fast. Almost always you are above clouds and storms.

And now to your inner growth work—

SHORT FOCUS IMAGING/AFFIRMATION

I image myself in this airplane in level flight.

I image myself relaxed, confident, and feeling secure.

I will decide then to unbuckle my seat belt or not, to get up and walk around or not, to look out the window or not, to eat and drink or not.

I develop and repeat my own AFFIRMATION or use one like this: I ENJOY THIS FLIGHT.

MYAFFIRMATION:________________________

____

____

LONG FOCUS IMAGING/AFFIRMATION

I image myself completing this flight successfully. I image myself leaving the airplane, walking the jetway, and entering the terminal building.

I develop my own AFFIRMATION or use one like this: I ARRIVE AT MY DESTINATION AFTER A SUCCESSFUL FLIGHT.

MYAFFIRMATION:________________________

____

____

PHASE FIVE: ACTUAL IN-FLIGHT: WOW!

The front part of the airplane seems to be coming down, the engine sound changes. We are in level flight.

TRANSITION FROM PHASE FOUR TO PHASE FIVE

I congratulate myself on the low anxiety I experienced during the take-off and the climb to level flight. I ACKNOWLEDGE THIS PROGRESS AND GROWTH IN ME.

SOME DECISIONS I MAY WISH TO MAKE:

Shall I unbuckle my seat belt or not; shall I get up and walk around or not; shall I eat or drink or not; shall I look out the window or not; shall I go into the restroom or not?

I enjoy these feelings of relaxation, confidence, and security. I thrill that I am speeding so fast to my destination.

I add up all these feelings, and I sense a great big WOW! THIS IS TRULY A "WOW" EXPERIENCE.

If I need to, it's OK to practice relaxation and/or deep breathing. I may read, or meditate, or talk to my travelling companion, or write, or nap, or . . .

A MYTH ABOUT THE SKY

1. I realize that the mediae of newspapers and television often do an injustice to reports about airplanes and air. It is a myth that air "comes and goes." In truth, the sky of

air is more a sea of air. Air is substance. Air has many of the qualities of water.

When an airplane suddenly goes up or down, there is nothing "wrong" with either the airplane or with the air. The airplane is simply moving with the air movement in the same way a boat moves with the water movement. Although the sudden up or down movement of an airplane is unnerving, there is rarely any danger since, in truth, the airplane is simply moving with the movement of the air.

2. It is an important truth that most severe storms are avoided. Occasionally, however, the air is bumpy in the vicinity of a severe storm. Lightning is usually frightening to passengers. Lightning almost always appears closer than it really is. And, if lightning were to strike an airplane, there is little to no chance of damage because the airplane is not grounded.

How much more fun and secure we are when the truth is known about air and lightning.

As you are flying along in this state of exhilaration of flight, you may wonder how you can be going so fast (550 miles an hour, often) but not sense the speed. The reason is that there is nothing up there with which to compare your movement. The closer you come to clouds, the more real sense of speed you have.

By the way, a fun game during flight is to look at the clouds and let your imagination loose: "I get hungry looking at all that cotton candy. I expect to see a big polar bear. Looks like skate or sled tracks. There is a huge castle."

This may be more than a "fun game." To focus on the clouds this way could help you look out the window, if looking out the window has been frightening to you.

THE AIRPLANE WINGS

There is a physical law that states correctly that an object that is flexible is less likely to break. This flexibility is consistently built into tall buildings, bridges, and poles; and you no doubt have noticed the "swaying" motion of such structures.

Aeronautical engineers use this same physical law, and they have designed the outer section of the wing of your airplane to be flexible. During flight, you may notice the outer part of the wing—primarily the wing tip section—moving up and down. Accept this as yet another certainty of the marvelous mechanical wonder and the security of your airplane.

MAKE A TELEPHONE CALL?

You have checked in your "Preparation at Home" section whether there will be a telephone on this airplane and exactly what you need to have with you and to do if you wish to call someone while you are flying.

There is a lot of value to contacting a person "down there" to share the <u>successes you are attaining at this point</u>. One option is for you to have cleared a person and a time range for someone to expect a call from you. Don't be disappointed about feeling super good that your anxiety about flying is reducing and your enjoyment of flying is increasing and have that person not be there to receive your call. A cautious note is to suggest to you whether or not you want to commit yourself in advance to a call. You might not welcome this additional pressure.

An attractive option is to discuss a possible call in advance with your special person. That person needs to know that you are working this program and be very familiar with what you are doing and why. The following is thus suggested. (1) Clear carefully that your friend will be available within a time range. You can compute a time range, allowing some time for the airplane to possibly take off a little after the scheduled departure time, to climb to the cruise altitude, and for you to practice your relaxation techniques, as may be needed. (2) You limit the time and content of your call. Talk primarily about the successes you are having. "______________________________, I am less anxious and more relaxed than I have been in an airplane in a long, long ('ever') time." Take a minute or so to describe the view out your window (if you are not too uncomfortable doing so). Remember, the purpose of this call at this time is to share your success on this flight and to receive a positive reaction from the special person to whom you are speaking.

And now to your inner growth work—

SHORT FOCUS IMAGING/AFFIRMATION

I image myself in this airplane, as we descend, prepared for landing, and then landed. I image myself relaxed, confident, and feeling secure.

I develop and repeat my own AFFIRMATION or use one like this: I THRILL TO THIS EXPERIENCE OF

FLIGHT. THIS IS TRULY A "WOW" EXPERIENCE!

MY AFFIRMATION: ____________________

____

____

LONG FOCUS IMAGING/AFFIRMATION

I image myself completing this flight successfully. I image myself departing the airplane, walking the jetway, and entering the terminal building.

I develop my own AFFIRMATION or use one like this: I ARRIVE AT MY DESTINATION AFTER A SUCCESSFUL FLIGHT.

MY AFFIRMATION: ____________________

____

____

PHASE SIX: PREPARATION FOR AND LANDING: KNOWLEDGE AND TRUST; OBSERVATION AND CONFIRMATION

Commercial airliners, like this one, now fly so high and so fast, the descending procedure for the landing usually begins way before the passengers expect it will start. The let-down procedure may start over a hundred miles from your destination and perhaps 10 to 15 minutes before passengers even think about landing.

Your first clue is that you may detect a change in the engine noise and possibly a slight change in the attitude of the airplane. Usually, no announcement is made by the pilots at the initial stage of letting-down, which is the descent to landing.

WHEN LANDING IS DELAYED. Sometime during the announcements regarding landing, there turns out to be an announcement that the landing will be delayed. Many pilots share this landing delay in a light, humorous way with such statements as, "We're going to go around in a few circles up here before we land." Or, "They want us to see some scenery up here before they clear us to a landing." Or, "Ladies and gentlemen, the traffic is heavy down there right now, so they want us to play around up here a bit before we go on down there to land." Around the busy airports, incoming jet airliners are required to slow down to a predetermined speed range and every airplane is placed in a precise position sequence for the approach and landing. Every airplane is in a precise spacing position for landing, so you can arrive at your destination in the most effective way and efficient time.

Two main factors impact on the delay of your landing—unusual heavy air traffic and less than perfect weather. A pilot will almost always announce if landing is delayed.

Remember, your airplane—and all the other airplanes near you and the airport—are being monitored by radio communication and watched by radar contact. Your flight crew will probably receive one of two possible instructions: either fly a distance beyond the airport, turn around, and return for the landing, or, be placed in a "holding pattern." Astute or experienced passengers will have a feeling that their airplane should be landing, and when that is not occurring, they understandably become somewhat anxious. Sensitive pilots will announce the landing delay; sometimes they will say how long they expect to remain in the landing delay.

If your pilot is expected to fly a course away from and then turn and fly back to the airport, you will have little or no sense of the airplane turning. You may simply realize the time delay in landing.

If your pilot is instructed to fly in a holding pattern, you may wish to note several items. A "holding pattern" is usually a very precise flight pattern that is "racetrack" in design. That racetrack pattern is always flown over an exact radio fix on the ground, and it is always flown at a rigid altitude.

Especially important to note is that most often more than one airplane is also in a holding pattern, the same holding pattern that you are in, except, that there is always a 1,000 foot separation between every airplane in the holding pattern.

When an airplane is cleared to land, the total holding pattern of airplanes literally and together moves down to the next one thousand foot level. This occurs because the lowest level airplane has landed.

You probably will not be aware when your airplane descends to the next lower level. You may notice one or more airplanes in the holding pattern (usually above or below you). This is absolutely no cause for alarm: the required distance separation is maintained at all times, and every airplane movement is being monitored by radio and observed by radar.

You quite conceivably could sense the airplane turning, either by your inner senses or the fact that the sun and/or shadows appear to slowly "move around" inside your airplane. Again, the exact holding pattern of flight dictates that the airplane flies in one direction a set time, makes a precise time turn, flies in that opposite direction a set time, makes a precise time turn, and so on. Your inner senses or the position of the sun and shadows, which seem to move inside the airplane, may cause you to become somewhat dizzy. We suggest that you either focus on some item, such as a book or magazine, on your lap or on the seat table or close your eyes. The turns, although precisely timed, are slow and at a minimum bank of the airplane. But if you are in the holding pattern long enough, you may sense the turning and feel a little "funny" or just a small amount of disorientation. All probably very minimal. This, too, shall pass.

Be patient, the requirement of orderliness, preciseness,

and sequence, are for the overall benefit of all the passengers in all the airplanes. Use this time as an opportunity to "prepare yourself for landing": work through the Short Focus Imaging/Affirmation and Long Focus Imaging/Affirmation (the end of this next to last phase of the program), deep breathing, and meditation. Hey, you are getting more than your money's worth by this extra time in the air.

As your airplane is approved to leave the holding pattern and then is cleared to land, the pilots will simply intersect the glide scope approach: regardless of the weather—clear, obscure, or in between—and regardless whether day or night, your flight crew will simply line up on the radio or light beam and follow that beam precisely to the airport and exactly to the end of the runway on which you will be landing.

The whole approach and landing are exciting, especially when you know what, why, and how it is taking place. And, you will.

TRANSITION FROM PHASE FIVE TO PHASE SIX

1. Congratulations for your anxiety reduction throughout all the phases up to this point. Ideally, while in level flight, you found yourself enjoying the flight. It is important for you to acknowledge your achievement. Any progress whether it consists of a sense of having fun flying or of getting up and walking around during the flight deserves that you recognize significant growth on your part. I ACKNOWLEDGE THIS PROGRESS AND GROWTH IN ME.

VERY NOTICEABLE CHANGES IN THE SOUNDS AND ATTITUDE OF THE AIRPLANE WILL BEGIN NOW.

RIGHT NOW, focus on relaxation and deep breathing. A suggested AFFIRMATION: I AFFIRM FEELING SECURE AS MY FRIEND THIS AIRPLANE CHANGES IN SOUNDS AND ATTITUDE.

From now until the landing is completed, you will be sharp, knowing what is happening and why it is happening. You will trust that everything being done is routine and that the airplane is responding—as always—to physical laws and that the pilots—as always—are following detailed checklists and are in constant radio contact and radar coverage. You will be observing what is happening, and what you see confirms what you expect to be happening. During this whole let-down and landing phase, you probably know more about what is happening and why it is happening than 90% of all the other passengers. This knowledge and trust, observation and confirmation will play a vital role in the reduction of your anxiety as your successful flight now nears completion.

THE LET-DOWN AND LANDING PROCEDURE

1. You will probably note a decrease in sound. Two factors contribute to this reduction in sound. The first factor is that engine power is being reduced. The airplane will be gliding through the air on its own rather than being pushed through the air by engine power. Your airplane is designed to glide many, many miles with no engine power at all. Mostly, the engine power will be reduced. That changes the sound. The second factor is that in order for

the airplane to glide while maintaining about the same speed, the nose may be lowered some. In some airplane models, the nose may be raised some. As the attitude of the airplane changes, the sound changes; that sound change is the sound of air passing over the airplane. Thus, when the sounds of the reduction of engine power and the lowering or raising of the nose in a glide combine, you may detect a decrease in sound. This is normal and routine.

2. Your airplane has been flying at a fast speed. As the airplane gets within a certain distance from the airport at which it will land, "speed laws" are in effect. The pilots must slow the airplane down to conform to the speed laws. If you look at approximately the middle of the wing, you will see a portion of the wing lift up. This is the dive brake or "spoiler." Dive brakes are used to slow the airplane down in flight and to slow the airplane when actually landed. These dive brakes may be raised and lowered numerous times in the earlier stages of the let-down procedure. You might sense that the airplane feels a little "rough" as the dive brakes are extended (up).

Expect now or soon, that a pilot will make the announcement to prepare for landing. This announcement is required by the FAA and is normal and routine.

A pilot with good public relation skills will not only announce that the airplane is beginning the descent to landing (and weather information at your destination city) but also will give <u>a time figure</u>, such as, "We expect to be on the ground in 18 minutes." Many passengers hear this announcement and process the message in two ways. One way is, "Wow, that guy/gal is good. We're a couple of hundred miles from the airport and 29,000 feet in the sky, and he/she can tell us in minutes when we will be on the ground!"

You could do the same thing . . . if you were in the cockpit (flight deck) and knew which gauges to look at. One gauge (avionic) tells you precisely how far you are from the airport, the other avionic shows how fast you are traveling over the ground. Simple. The computation of these two figures results in the time to arrive at the airport. This is referred to as "ETA," that is, "estimated time of arrival." You are almost a co-pilot already.

The second part of processing the announcement goes something like this: "Hum. It is now 8:40. We will be on the ground in 18 minutes. 8:40 plus 18—that's 8:58." We'll be on the ground at 8:58. That magical 8:58 becomes etched in your mind, and your watch becomes a major attention grabber.

Caution. As the airplane gets closer and closer to the airport, the pilots must invest even more concentration on the airplane and its movements. Numerous things could happen on the ground or in the air near the airport after you receive the expected landing time: traffic could become heavier; weather could change.

Any change at or near the airport could cause an additional delay in your landing. The concentration of the pilots on the descent and landing understandably could prevent them from announcing the change in landing time. Also, after a 3 hour flight, 10 minutes isn't very much. To them, that is. To you, a 10 minute lapse from the announced landing time could—understandably—become a very big thing.

So, please, it's OK to compute the anticipated landing time. However, don't hold that figure as absolute or sacred. The lapse in time after your computed landing time in no way indicates a problem.

Instead, attend to the following items as you identify what and why everything is happening.

3. With the use of dive brakes and the eventual use of flaps being extended and the landing gear coming down, from now until the airplane arrives at the gate, your airplane is considered "dirty."

4. Expect that the engine noise will vary—increase, decrease, etc. As you depress, let up, depress, etc., the accelerator on your car or truck, so the pilots are adding and reducing engine power as needed. This is normal and routine.

5. A change in altitude causes some people to experience some discomfort (and in some cases, pain). You need to attend to this in advance of the airplane starting its descent, if this has been or is now a problem to you.

A change in altitude may cause some problems with your ears. Your airplane is equipped so the descent has minimal, if any, effect on most passengers. This problem is a physiological response to pressure changes.

If you are sensitive to pressure changes, the following is suggested. Stay awake during the descent and follow one or more of the following techniques: allow mints or hard candy to melt in your mouth; yawn enthusiastically; repeat swallowing; hold your nostrils and "push" air. This latter technique is encouraged and is described by the American Academy of Otolaryngology* in this way—

> "(1) Pinch your nostrils shut. (2) Take a mouthful of air. (3) Using your cheek and throat muscles, force the air into the back of your nose as if you were trying to blow your thumb and fingers off your nostrils. When you hear a loud pop in your ears, you have succeeded. You may have to repeat this several times during descent."

* The pamphlet, "Ears, Altitude and Airplane Travel", from which this quote is taken is available from American Academy of Otolaryngology, One Prince Street, Alexandria, VA 22314.

Three final notes for this section follow.

First, the airplane starts a descent many miles and many minutes away from your destination. Alert a flight attendant that your ears are very sensitive to pressure changes with altitude changes, and it is <u>important</u> for you to be informed when the airplane will be beginning to descend.

Second, if you have a baby with you, a baby cannot use the ear popping technique as you are able to do. Be sure, then, that your baby is awake and sucking on a bottle or a pacifier or eating.

Third and finally, almost always the use of one or more of the techniques will minimize or even stop the effects in your ears if pressure changes. If they do not, however, the descent will only continue for some minutes. Your discomfort will pass with time.

6. Not only will your airplane be coming down in altitude, it will probably be making several turns. Your airplane has been following precise radio "highways" all during your flight. A radio highway is a precise radio beam; a radio beam provides exact location and precise direction. The airplane will be making turns now as it joins a particular radio highway that will guide the airplane precisely to your airport of landing and exactly to the runway on which you will be landing. These altitude changes and turns are normal and routine.

7. You may hear a "churning" or "cranking" sound. (Remember these sounds from the airplane taking off?) This will probably occur several times as the airplane is landing. You can confirm what is happening: look out the window. See that "part of the wing"—flaps—are extending out from the leading edge of the wing and extending out and down from the trailing edge of the wing. Again, this changes the configuration of the wing so lift continues even though the airplane is slowing down a great deal.

You will likely note—as the airplane becomes more and more "dirty"—flaps extended and the landing gear down—that the airplane seems less stable—it may even seem to be wobbling some. This is because at slower speeds, the airplane does make more changes, which are normal and routine; near the ground the air becomes a little "rougher" as it bounces over trees, buildings, and so on. Although the noises and altitude changes make the airplane seem less stable, it is not less stable. You react to the noises and altitude changes by noting that all of this is normal and routine.

8. It is common for the airplane to be landing, and you know it is landing, but you can't see outside because it is dark of night or because of clouds. This is not a problem.

Your pilots are watching a gauge with two pointers or needles that looks like the drawing below.

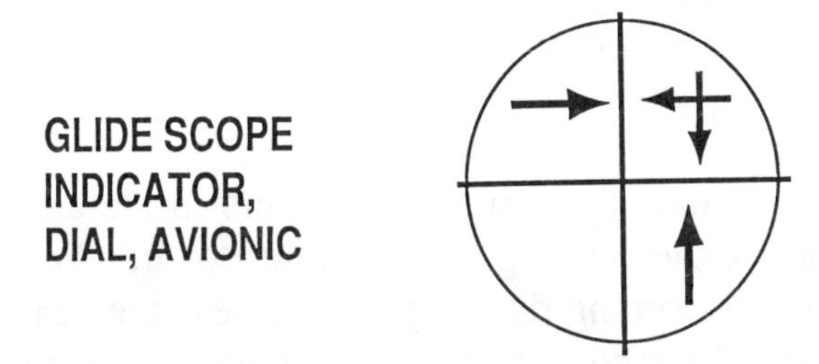

As the pilots keep these two needles centered, the airplane is following a radio beam that will bring your airplane exactly to the landing point on the exact runway of your destination airport.

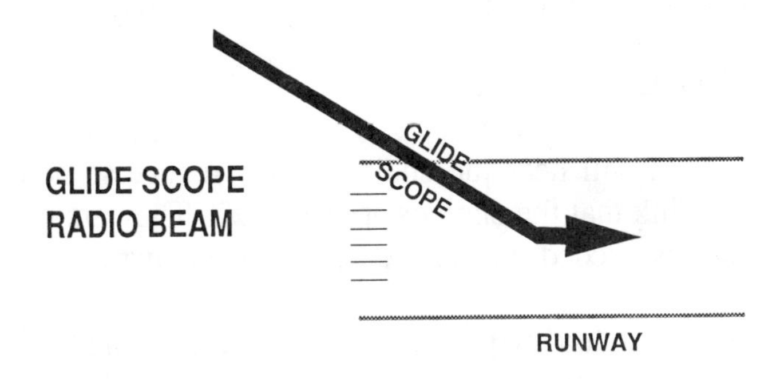

That radio beam and the precise way it works is called the "glide scope." The pilots do not need to be able to see outside the cockpit. They use this dependable instrument for an exact approach and precise landing. The landing is also being monitored by radar coverage. The radar observer could make suggestions to the pilots, if needed, which it rarely is.

Feel secure and confident that the whole let-down (descent) procedure—from decreasing engine power to turning off the runway after the landing—is normal and routine.

9. The final sound and shake while in the air will come to you as the landing gears come down. When all the wheels are coming down, you will hear the "churning" or "cranking" sound until they are down and locked securely in place. Now the airplane may feel especially wobbly and seem to be less stable. Such is not the case. You are now ready to land.

Relax . . . deep breathing . . . I AM TRUSTFUL OF THIS AIRPLANE, ALL THE WONDERFUL RADIO AND RADAR AIDS, AND THE PILOTS.

THE LANDING

1. You will hear the engine power reduce so much you may think that the engines are shut off. They are not. NO power is needed during the active touchdown.

The sound may seem to decrease a lot and you may note the front of the airplane raised a little. This is basically the same configuration, or attitude, the airplane was in during the rotation on take-off. The pilots will maintain this attitude until the airplane is no longer flying and it touches the runway.

Most landings are so smooth, passengers are uncertain exactly when the airplane quit flying and touched the runway. There are many factors that could—on rare occasions—cause the airplane to bounce just a little in landing. There is an aviation joke: A passenger is leaving the airplane at the gate. The passenger looks into the cockpit or flight deck and says, "That was a great landing. The co-pilot must have been flying the airplane."

You might detect that either the left or right wing is low just before the airplane settles on the runway. This will occasionally be the case; the pilot is flying the airplane purposefully that way on landing. A low wing simply means that the wind is blowing from that direction. This is an approved and taught landing technique. Incidentally, with a strong "cross-wind", the wheels on the low wing side of the airplane may touch down first, then the wheels on the other side of the airplane. This feels like a "bumpy landing." In fact, it is an excellent landing, considering the wind direction.

2. Just as soon as the airplane is rolling on the runway, you will hear a sharp increase in sound and a definite "tug" on your body. As soon as the airplane is rolling on the runway, you will hear a sharp increase in engine noise. The engines are in "reverse" and are running fast simply to slow down the airplane. Out on the wing you may notice the dive brakes come up. The pilots are using the brakes some. All of these combine to push you forward in your seat belt. Again, the landing is normal and routine.

3. The airplane will now turn off the runway and taxi to the gate.

PERSONAL OPTION: If you feel the need to get off the airplane as quickly as possible, call a cabin attendant and indicate that need to her or him.

And now to your inner work–

SHORT FOCUS IMAGING/AFFIRMATION: I image myself leaving the airplane, walking the jetway, and entering the terminal building. I image myself relaxed, confident, and feeling secure.

Develop and repeat your own AFFIRMATION or use one like this: I GRATEFULLY AFFIRM MY SENSE OF SECURITY AS THIS FLIGHT IS SUCCESSFULLY COMPLETED.

MY AFFIRMATION: ____________________

____

____

LONG FOCUS IMAGING/AFFIRMATION: I image myself completing this flight successfully. I image myself departing the airplane, walking the jetway, and entering the terminal building.

This AFFIRMATION or your own: I ARRIVE AT MY DESTINATION AFTER A SUCCESSFUL FLIGHT.

MY AFFIRMATION: ____________________

____

____

PHASE SEVEN: TAXIING TO THE GATE; LEAVING THE AIRPLANE: I DID IT! GREAT JOB! THANKS!

As the airplane turns off the runway and begins taxiing to the gate, this is an opportune time to REALLY CONGRATULATE YOURSELF. Any reduction in your anxiety at any point during the flight deserves your self-congratulation. I ACKNOWLEDGE THIS PROGRESS AND GROWTH IN ME.

Unless you feel an impelling need to leave the airplane quickly, it is suggested that you remain in your seat while all or most of the passengers are lining up and then hustling and bustling to leave.

FOCUS ON YOUR GOOD FEELING OF BEING RELAXED, CONFIDENT, AND SECURE.

And now your final inner work—

SHORT FOCUS IMAGING/AFFIRMATION

I image myself leaving the airplane, walking the jetway, and entering the terminal building. I image myself relaxed, confident, and secure. I especially image myself as smiling and being happy.

I develop and repeat my own AFFIRMATION or use one very similar to this one: I HAVE ARRIVED AT MY DESTINATION AFTER A SUCCESSFUL FLIGHT!

I develop and repeat my own AFFIRMATION or use one like this: I AM GRATEFUL FOR THE PROGRESS AND GROWTH THAT I MADE ALL DURING THIS

AIRPLANE TRIP. I DID IT! GREAT JOB!

MY AFFIRMATION: ______________________
______________________________________!

An appropriate action is to thank the cabin attendants as you pass them, and look into the cockpit or flight deck and thank the flight crew. The cabin attendants and flight crew are human beings just like you. They deserve—and appreciate—hearing compliments from you.

NOW, PAUSE BRIEFLY AT THE DOOR OF THE AIRPLANE AS YOU EXIT.

AFFIRMATION: I BLESS THIS AIRPLANE AND ITS CREW FOR MY SUCCESSFUL FLIGHT.

(TO THE AIRPLANE) THANK YOU, MY FRIEND!

FINALLY WALK THE JETWAY AND GO INTO THE TERMINAL BUILDING—SMILING AND HAPPY.

THE DEVELOPERS' ULTIMATE "REWARD"

All the time and effort we have invested in this program is rewarded by your successful completion of the program. Our reward is that each user of this program will experience a reduction in anxiety about flying and an increase in enjoyment of flying.

There is an ultimate reward to us: THAT YOUR SPIRITUAL BELIEFS HAVE BEEN EXPANDED AND YOUR SPIRITUAL FAITH DEEPENED. No other reward will mean more to us than that. We affirm that experience for you.

Jim and Leona Remington

AND SO WE SHARE WITH YOU . . .

Many people consider that you are soon to begin the "dangerous part" of your whole trip: your drive or ride home.

We bow out of your success in this flight and your active involvement in the program, by congratulating you and by saying to you

. . . PLEASE DRIVE CAREFULLY

. . . FLY ENJOYABLY

. . . AND, THE NEXT TIME YOU MAKE A TRIP BY AIR, FLY RELAXED, FLY CONFIDENT, AND FLY FEELING SECURE

. . . WITH OUR BLESSING, SOAR LIKE THE EAGLES

OUR THRILL WITH YOU—THE USER OF THIS PROGRAM

With the reduction of fear and anxiety about flying, and with the resultant increase of enjoyment of flying, we experience "THRILL."

As is very apparent to you as you have worked this program, we the developers thrill by riding on a commercial jet airliner. And we wish that same thrill for you.

We thrill with you by quoting the following poem:

"Oh! I have slipped the surly bonds of Earth
And danced the skies on laughter—silvered wings;
Sunward I've climbed, and joined the tumbling mirth
Of sun-split clouds."

-John Gillespie McGee, Jr.

APPENDIX A: HELPFUL ITEMS AND INFORMATION

BOOKS

The following books are either available for ordering and purchasing now or may be ordered with delivery date to be announced. Please see page 113 for the Information Request Form.

Avioanxiety Becomes Controlled: Now, Fly Without Fear (C). The same basic book/program as this one, except that it is in cassette form. An ideal way to deal with anxiety and fear, listening while driving, or relaxing in your house, motel, or hotel room.

Avioanxiety Becomes Controlled: Riding on a Corporate Aircraft. The contents of this book present a systematic procedure of anxiety and fear reduction for the executive who must travel on business airplanes.

I Am Cleared For Take-Off. A fantastically interesting and helpful book that moves the reader through all the important subjects connected with flying by commercial jet airliner. Many of the subjects are presented with a personal commentary by Dr. Remington; there is a great deal of perceptive insight and humor involved. After reading the book, the reader feels exceptionally knowledgeable about all aspects of commercial aviation and feels that he/she has just completed a thrilling flight on a commercial jet airliner. I Am Cleared For Take-Off is an excellent complement to the book you are now reading.

Spiritual Truths That Fly. A series of devotional/inspirational readings. This uplifting book is unique in that it may be the only book that relates Biblical texts and spiritual truths to some aspect of aviation. Each reading contains an appropriate meditation. This book has especial appeal to persons who hold spiritual values and have an interest in aviation.

This Time.. I Think, I'll Fly. A brief, but thorough, introduction to flying the first time on a commercial jet airliner. A systematic order moves the reader with familiarizing information and practical ideas, beginning at home and continuing to the completion of the flight. Relaxation techniques are suggested. The book is carefully designed to assist the reader to experience an enjoyable first flight on a commercial jet airliner.

CASSETTE TAPE

"Empowerment: in Quietness and Confidence Shall Be Your Strength" offers the listener a process of relaxation and guided meditation to improve the quality of living. There are four helping/guiding segments: Preparing For the New Day (morning); Turning Hurts Into Healings; Growth From Mid-Day Stressors; Welcoming the Night. Background sound is the ocean.

GAME

"Windsock." An interesting and educational game that increases an understanding and appreciation of the exciting world of commercial aviation, including riding on a commercial jet airliner. This family-type game is designed for persons ages 8 through 108. Any number of persons may play this game.

BOOKLETS

"Stress Management: It's In Your Pocket." A novel and practical guide to identify the source and impacts of stress with specific techniques of managing that stress. Of particular value is that a "stress indicator" is included.

"Doing It For the First Time: What Both of You Should Do." This material provides specific guidelines to making an individual's first airplane ride in a small, general aviation airplane a fun experience. One section applies to the passenger; the other section applies to the pilot. An attractive "First Flight" Certificate is included.

"When and Where? The First Desire of the Human Being to Fly." A fascinating little booklet that researches mythology to discover when human beings began to wish that they could leave the ground for a controlled and extended length of time. This item is an excellent source to use before looking at the traditional history of flying.

SPEAKING

The developers of this program may be engaged as speakers, entertaining and informative, in a variety of settings and on a wide spectrum of non-aviation and aviation topics. A list of topics is available. Also, this couple works as a team in making presentations during church and chapel worship services.

COUNSELING

Dr. Remington is in private practice in St. Louis and offers an especial caring and sensitive approach with his professional therapy services. He may be contacted directly to inquire about his training, degrees, license, experience, and fees, or he will be happy to send you a brochure detailing his services. He provides a wide-range counseling service to help the client deal more effectively with issues that impact on important life matters and the quality of life and living.

CONTACT PROCEDURES

Telephone contact: 314-533-4150, or send the following Request For Information form.

INFORMATION REQUEST FORM

Jim/Le,
Please send the followed checked information to me:

<u>books</u>

_____ Avioanxiety Becomes Controlled: Now, Fly Without Fear (C)
_____ Avioanxiety Becomes Controlled: Riding on a Corporate Aircraft
_____ I Am Cleared For Take-Off
_____ Spiritual Truths That Fly

<u>cassette tape</u>

_____ Empowerment: in Quietness and Confidence Shall Be Your Strength

<u>game</u>

_____ Windsock

<u>booklets</u>

_____ Stress Management: It's In Your Pocket
_____ This Time .. I Think, I'll Fly
_____ Doing It For the First Time
_____ When and Where?

<u>speaking</u>

_____ List of topics and occasions of your personal appearance and speaking
_____ Information regarding your counseling services

MAIL THIS FORM TO:
Inner Marker to Growth, Attention: Dr. Jim and Leona Remington, P.O. Box 23310, St. Louis, MO 63156, U.S.A.

APPENDIX B
FORMS OF TRANSPORTATION - A COMPARISON

TRIP:______________________________

NOTE: Be sure that all computations and evaluations are based on a round trip. Also note, compute and evaluate from the perspective that if you decide to make the trip by air, your anxiety—in comparison to other trips you have made by air -will on this trip be reduced, and your enjoyment of flying will be increased. Mark those areas that actually apply to you.

MAJOR COSTS:	AIR	PERSONAL CAR	RENTED CAR	VAN	TRAIN	BUS	BOAT
Meals (plus tips)							
Motel/Hotel							
Parking							
Repair/Towing							
Estimated wear and tear on vehicle as cost to you							
Gas/Oil							
Other							
TOTAL COST							

TIME:	AIR	PERSONAL CAR	RENTED CAR	VAN	TRAIN	BUS	BOAT
To eat							
To use restrooms							
Overnight							
For repairs/towing							
Detours							
Other							
TOTAL COST							
RELAXATION: (Score "more," "some," "lots")							
WEAR AND TEAR ON YOU—PHYSICAL (FATIGUE FACTOR): Yawning; struggling to keep awake; blurring of vision; an overall sense of physical "numbness." (Score "low," "medium," "high")							

	AIR	PERSONAL CAR	RENTED CAR	VAN	TRAIN	BUS	BOAT
WEAR AND TEAR ON YOU - MENTAL (ALERT FACTOR): Paying proper attention; responsiveness; quick decision making; an overall sense of mental "numbness." (Score "low," "medium," "high")							
WORK PRODUCTIVITY—CAN WORK WHILE TRAVELING (Score "yes," "no")							
PROBLEMS WITH TRAVELING WITH FAMILY: (Score "more," "few," "many")							
POSSIBLE INTERFERENCE FROM THE WEATHER: (Score "no," "?" "yes")							
POSSIBLE DELAY EN ROUTE: (Score "long," "short," "?" "more")							

	AIR	PERSONAL CAR	RENTED CAR	VAN	TRAIN	BUS	BOAT
INCREASED TIME AVAILABLE FOR YOU AT YOUR DESTINATION: (Score "more," "some," "lots")							
AGGRAVATION WHILE TRAVELING—IMPATIENCE, LOSS OF HUMOR, DISAPPEARANCE OF FUN: (Score "increased," "?" "decreased")							
OVERALL CONVENIENCE (Score "most," "medium," "least")							
OVERALL ALERTNESS/RESPONSIVENESS UPON ARRIVAL AT DESTINATION: (Score "dull," "neutral," "sharp")							
FINAL COMPARISON RANKING							

OVERALL FINAL COMPARISON - SEE NEXT PAGE

GO BACK AND REVIEW EACH ITEM CAREFULLY. DO NOT MAKE A "SWEEPING EVALUATION" BECAUSE DIFFERENT RESPONSE "SCORES" HAVE DIFFERENT POSITIVE/NEGATIVE VALUES. FOR EXAMPLE: A "HIGH" SCORE ON WEAR AND TEAR ON YOU—PHYSICAL AND MENTAL—IS A VERY NEGATIVE VALUE.

FINAL COMPARISON RANKING: SUM UP—ACTUAL DOLLARS AND TIME COMPUTATIONS AND THE "EMOTIONAL" VALUES OF ALL OF THE ITEMS.

FILL IN THE "FINAL COMPARISON RANKING" (PREVIOUS PAGE) IN THIS WAY:

HIGHEST POSITIVE RANKING "1"
NEXT DESIRABLE "2"
INCREASINGLY LESS DESIRABLE RANKINGS "3," "4," "5," "6," AND "7."

FOCUS ON YOUR HIGHEST RANKING ONE(S) AS PROVIDING YOUR HIGHEST MOTIVATION TO TRAVEL.

AVIATION GLOSSARY

- A -

ACTIVE RUNWAY. At every airport one runway is designated as the "active runway." This designation defines the runway number and determines the direction from which airplanes take off and land on that runway.

AERODYNAMIC FORCES. There are four fundamental forces also called aerodynamic forces or physical laws, operating on the flight characteristics of an airplane: thrust, lift, drag, and weight. The physical laws are that the airplane flies when lift is greater than weight and thrust is greater than drag.

AERONAUTICAL KNOWLEDGE. One of the many strict requirements pilots of commercial airliners must meet before they are approved as pilots. Airline pilots must know a great deal about the whole field of aviation, commercial aviation in particular; they must have logged a minimum amount of flying time, including flying time in their respective airplanes, and they must be able to perform numerous basic and precision flying skills, especially flying skills in their respective airplane. Pilots are required to pass demanding physical examinations every six months and to pass flying proficiency tests on a regular basis.

AFFIRMATIONS. The stating and the repeating of spiritual truths. Affirmations bring up from deep within a person the most profound and practical truths of who that person really is and what that person is really capable of doing. Affirmations provide the opportunity to take advantage of spiritual truths.

AGE, AIRPLANE'S. It is inaccurate to think of the age of an airplane in terms like "it's rusting out" or "when it breaks, I'll fix it." FAA rules and regulations require an overhaul of a part by "time in service," not when it needs fixing. This applies to all the mechanically important parts on the whole airplane. In a literal sense, an airplane does not "get old."

AILERONS. A mechanical part located from the middle to outer wing tip, on the trailing edge section of the wing, manipulated by the pilots on the flight deck to roll/bank the airplane to the right or to the left, along the longitudinal axis.

AIR. A combination of a gaseous mixture and oxygen. Air has definite properties and is understood by consistent physical laws. The movement of an airplane through the air, then, is completely predictable. Air moving is called "wind."

AIR TRAFFIC CONTROL (ATC) Every part of the flight of a commercial jet airliner is under strict rules and regulations. There is constant radio communication with and radar coverage of every flight.

AIRPLANE. An engine-attached, fixed-wing flying device that is heavier than air and supported in flight by the flow of air around the wing.

AIRPORT. An area on land or water from which aircraft take off and land and usually contains buildings and aviation facilities. A visit to an airport opens up a whole world of experiences.

AIRPORT DESIGNATED OBSERVATION AREA. A location on or near an airport that provides a clear view of the airport's exciting activities.

AIRPORT LANDING SYSTEM (ALS). Modern airports are equipped with a variety of aids pilots may use when landing. Airplanes landing in limited visibility use highly sophisticated landing aids, and the landing becomes routine. These aids may be special lights and lighting, electronic waves, or radio beams.

AIRPORT VISIT. Some people benefit from a desensitization schedule that allows them to arrive at the airport with less anxiety. A familiarization visit to your airport prior to an actual flight also assists in decreasing the anxiety associated with a flight.

AISLE SEAT. Aisle seats are preferred by some passengers because the passengers feel less confined and experience less sensation of the airplane banking.

ALTERNATE AIRPORT. Before a commercial airliner leaves its departure airport, an alternate destination airport has been selected by the pilots. If it is not possible to land at the destination airport the pilots simply fly to the previously selected alternate airport and land.

ANXIETY. The awareness of a sensation that may range from a barely discernible and vague uneasiness all the way into the early stages of intense fear. An extreme case of anxiety may result in an anxiety attack. An anxiety attack shows itself with physical symptoms, thus, a correct diagnosis of anxiety attack is often not made. A diagnosis of a physical problem of some kind is often mistakenly made.

APPROACH CONTROL. A facility that gives directions for and monitoring of the descent and approach-to-landing of commercial airliners.

AUTO-PILOT. A highly sophisticated avionic located on the flight deck that literally flies the airplane. Pilots give specific commands to the auto-pilot, and those commands are translated into the movement of the airplane's controls.

AVIATION. One of the most thrilling and valuable accomplishments in the history of humankind. Just imagine! The ability to overcome the pull of gravity and to soar like the eagles! To fly reflects one of early humankind's deepest desires.

AVIOANXIETY. This concept combines the feeling and reality of anxiety with the whole field of aviation. The word was professionally researched by the Remingtons and was determined not to exist at this time. The word is copyrighted by the Remingtons and is not available to be used by anyone else.

AVIONIC. A high-technology device on an aircraft that assists the pilots in one or more aspects of the airliner's total flight.

AXES. There are three distinct axes that impact on an airplane. The <u>longitudinal axis</u> runs the length of the airplane. The airplane banks or rolls around this axis. The pitch axis runs across the airplane. The airplane goes up or down around this axis. The yaw axis runs vertically down through the middle of the airplane. The airplane turns around this axis.

- B -

BACK-UP SYSTEMS. Unlike the vehicles with which we are probably familiar, the airplane has at least one back-up system to every system. If for any reason one system does not function as it is designed to do, the back-up system replaces the operation of that system.

BEHAVIOR THERAPY. Most people act and react based on what they receive as rewards or negative consequences. This program emphasizes rewards and positive reinforcements.

BRAKES. Every airplane has brakes, like other vehicles. Airplanes, unlike other vehicles, can also reverse their engines in conjunction with applying the brakes when on the ground after landing.

BREATHING EXERCISES. The use of deliberate deep breathing exercises provides valuable physiological and psychological benefits to a person. Deep breathing exercises are important in this program.

- C -

CABIN PREPARATION. Clearly defined and very specific rules and regulations are imposed by the FAA as the airplane is preparing to take-off and is preparing to land. The flight attendants communicate this vital information to the passengers by the making of announcements, demonstration, and then by personal contact. All passengers should give the flight attendants' announcements and demonstrations careful attention.

CALMING RESPONSE. The internal human mechanism that prepares the body to respond to a perceived thrust can also provide a calming response. This program is designed to send signals to the body to be calm regarding anxiety and fear. The technical name for this calming response is the "parasympathetic response."

CARRY-ON LUGGAGE. A certain amount and size of luggage may be carried on board the airplane, rather than checked in as baggage.

CHAPEL. More and more airports are providing, or approving, chapels. Many passengers, flight crews, and visitors, appreciate a quiet place reserved for individual or small group meditation and prayer as well as a sacred place to conduct religious services. Most chapels, by design and decor, enhance and fulfill the spiritual needs of various faiths.

CHAPLAINS; CHAPLAINCY PROGRAMS. More and more airports have chaplains and chaplaincy programs. Chaplains are in the airport or are on 24-hour call. Chaplains are usually well-trained; they serve the needs of their respective faith, and serve people of any faith or no formal faith. Chaplaincy programs usually reflect this same interfaith approach. Most airlines are very cooperative to contact a chaplain on the request of a passenger.

CHECK-IN. Passengers who have not purchased their tickets in advance check in at the airline ticket counter. Passengers who have purchased their tickets in advance, depending upon the size of the airport and other factors, may check-in at their respective airline curbside counter or in their respective airline waiting area.

CHECKLIST. A checklist is a "sacred document" to pilots. Pilots are <u>absolutely</u> <u>required</u> to follow in exact detail and precise order a checklist for every facet of their flight throughout the flight.

CHURNING/CRANKING SOUNDS. Anxieties and fears are reduced when sounds are expected and the meaning and purpose of those sounds are clearly understood. This program does exactly that for its user.

CLEAN. An aviation term denoting that an airplane's flaps and landing gear are all retracted. The airplane is as streamlined as possible.

CLOTHES. This program encourages passengers to select and wear clothes that are casual and loose-fitting, especially around the waist and neck.

CLOUDS. Clouds indicate an accumulation of fine droplets of water and/or very small particles of ice. There are "clouds" of steam or haze or dust or smoke. Clouds are "signposts in the sky" to pilots. A good pilot can predict what is happening weatherwise by studying the clouds.

COGNITIVE THERAPY. An approach in which people receive information designed to change their ways of thinking. This program provides extensive information to assist in this process.

COMMUNICATION. An airliner in flight is in constant communication with numerous facilities during the course of a flight. Airplane pilots frequently exchange information with pilots flying other aircraft. The total flight of an airliner is being followed with direct contact by radar and radio operators and often by other airplanes.

CONDITIONS OF LEARNING. This book/program is essentially an educational program. The learning philosophy undergirding this educational program is that the best learning takes place when the external and internal conditions of learning are optimal. This philosophy of learning is carefully interwoven into and exists throughout the program. The major learning philosophy was borrowed from the adult educator Robert Gagne'.

CONFIDENCE BUILDING. An individual experiencing some anxiety or fear in at least one area of her/his life often is low in self-confidence. This program deliberately builds self-confidence in the user as the user proceeds in working the program as it is written.

- D -

DE-ICING or DE-SNOWING AIRPLANES AND RUNWAYS. Before an airplane leaves the gate, if there is ice or snow on the airplane or if ice or snow is forecast anywhere along the flight route, the airplane is sprayed with an anti-icing fluid. If ice is encountered while in flight, the airplane has mechanical means of causing the ice to chip and then blow off the surface of the airplane.

DELAYS. With the popularity of air travel understandably increasing, it is inevitable that passengers will encounter delays, delays in leaving the gate, delays in actual take-off, and delays in landing. No one wants to complete your flight more than do the flight crew. Delays are not of the making of the flight crew; the flight crew cannot have any control over the delays. BE PATIENT; BE PATIENT; BE PATIENT.

DESCENT. Because commercial airliners fly so high and so fast, the descent begins before most passengers expect and are prepared for it. Knowing when the descent is going to begin is important to passengers who have problems with their "ears popping" because of the pressure changes as the airplane descends. Pressure change techniques should begin at least during the start of the descent if not before the descent begins. People who have trouble with their ears popping should ask a flight attendant to let them know just before the descent is to begin.

DESENSITIZATION; DESENSITIZATION SCHEDULE. Some people's fear of flying is so severe, a phobia, they cannot even go to the airport without experiencing anxiety attacks. A slow, gradual, approach of going to the airport and confronting increasingly amounts of the object of one's fear, following a sensitization schedule, is often recommended. Desensitization is gradually approaching what has been a terribly negative experience.

DESTINATION; DESTINATION AIRPORT. The point of arrival; the place to which one intends to go; the end of the flight.

DIRTY. An aviation term denoting that an airplane's flaps and landing gear are extended (down). This configuration occurs just after an airplane has taken off and on its approach-to-landing and landing.

DIVE BRAKES. Mechanical wing-like devices, located on the wing. These are activated, raised and lowered, by the pilots to slow down the airplane. Dive brakes are used to slow down the airplane in making the transition from cruise to descent, during the descent, and as soon as the airplane is on the runway after landing. Also called "spoilers." They "spoil" the smooth flow of air over the wing that decreases lift and provides some resistance to the forward movement of the airplane.

DISTANCE MEASURING EQUIPMENT (DME). An avionic on all commercial airliners that displays the remaining distance to a selected point, usually the destination airport. That distance figure is computed into an estimated time of arrival, known as "ETA". Pilots often share this figure with the passengers.

DRAG. Drag is one of the aerodynamic forces, fundamental forces, a physical law, that must be overpowered in order for an airplane to fly. There are several "drags" on an airplane, physical properties that attempt to prevent the airplane moving swiftly through the air. Thrust must be stronger than drag in order for an airplane to fly.

DUAL AVIONICS. Commercial airliners always have at least two pilots. A pilot could fly the airplane alone. There are complete and identical sets of avionics in front of each pilot, two complete sets of controls, and all levers and switches essential to engine performance and other operating parts of the airplane are equally accessible to both pilots.

- E -

ELEVATORS. A mechanical part located on the horizontal stabilizer, the little wing on the tail section, that provides movement of the airplane up and down, along the pitch axis.

ENGINES REVERSE. Commercial jet airliners have the capability to be slowed down, in some cases to back up while on the ground, because the jet engine may be able to function in reverse or a mechanical part comes over the jet engine so the jet thrust is in reverse.

EXITS; EXIT DIAGRAM. All the exits on an airplane are clearly marked. Aisle lights will direct you to exits if it is dark inside the airplane. An exit diagram must be in the seat pocket in front of each passenger. The exit diagram should be carefully noted, especially in relationship to your seat and closest exit or exits to you.

- F -

FEAR. A powerful and intense feeling that is usually directed to an identifiable "target." A fear, however seeming to be unfounded, without basis, or "silly," to others, is very real to the person experiencing that fear.

FLAPS. Mechanical devices, located on the trailing edge and the leading edge of the wing. When the flaps are extended, that is, "down," the airplane is flown with total control and stability at a slower speed. The flaps are always extended on take-off and on landing. Flaps give the appearance that a part of the wing has slid forward and down (leading edge) and back and down (trailing edge).

FLIGHT. One of humankind's fantastic accomplishments has been to design and fly a vehicle that defies gravity and overcomes the pull of drag.

FLIGHT ATTENDANTS. Important members of the flight crew that primarily attend to the needs of the passengers and whose service area is the airplane's cabin. The older names for this key role are "stewards" and "stewardesses."

FLIGHT CREW. A generic term that may be applied to all persons involved in the flight in the air of a commercial jet airliner. Technically, perhaps the term should be limited to the personnel on the flight deck who manipulate the controls of the airplane.

FLIGHT DECK. More modern term for "cockpit." The flight deck is located to the front and top of commercial airliners. All the controls and all the personnel connected with the flight operation of the airplane are located on the flight deck. Passengers may look in or go in with the permission of a pilot before and after flights. Passengers are not allowed on the flight deck while the airplane is in flight. We encourage passengers to visit the flight deck and to talk with either one or more of the pilots, especially after a flight has been completed.

FLIGHT PLAN. Every detail of a proposed flight

is carefully planned and submitted to the appropriate FAA facility. That plan may be accepted as presented or modified by the FAA. The final flight plan is given to the pilots immediately preceding take-off, and that plan must be flown as given to the pilots and accepted by them.

FLOTATION DEVICE. Seat cushions on commercial jet airliners may serve to keep a person floating in water.

FOOD. The request for special food, because of medical or religious or philosophical reasons, should be made in advance of a flight and the request will be honored.

FRIEND. This book/program encourages and guides users of this program to make friends with the airplane on which the passengers will be traveling.

FUEL. The fuel in jet engines is kerosene. Passengers might smell kerosene when arriving at the airport, while walking the jetway to or from the airplane, or while on the airplane before the doors are closed to begin the flight.

FUSELAGE. The main "body" of an airplane. The fuselage houses the passenger cabin, flight deck, and baggage compartment.

- G -

GAGNE', ROBERT M. His philosophy of adult education is incorporated in this book/program.

GATE. A location in a terminal building or on a concourse at which there is a waiting area, a check-in counter, and a door leading to a jetway to an airplane. Each gate is identified by a number.

GLIDE SCOPE. An avionic that displays the precise approach path to guide the airplane to the predetermined landing spot on the active runway.

GO-AROUND. A maneuver practiced by pilots of commercial airliners. In any event where a landing is not proceeding as the pilots want, the landing is aborted, and the airplane continues in flight, usually circling the airport to begin another landing approach.

GROWTH BEYOND THIS PROGRAM. Based on the people who tested this program and those who have used it, it has become evident that as people reduced their anxiety about flying, they also became less anxious in other parts of their lives; as their enjoyment of flying increased, their enjoyment increased in other areas of their lives.

- H -

HABIT OF NEGATIVITY. Some people are very negative in life and about life. Some people have been this way for a long time. They have developed a habit of being negative. This habit needs to be broken or replaced by a habit of positiveness.

HABIT OF POSITIVENESS. We control the habits we assume in life. We can assume a habit of positiveness. It is remarkable how life becomes more satisfying and fulfilling when we approach life with a positive attitude.

HIGHER POWER. Many people find this an appropriate term for the Deity in their lives.

HOLDING PATTERN. When an airliner has been advised that there will be a delay in landing, the pilots are instructed to fly a holding pattern. This is usually a "racetrack" design over a specific location, or radio fix.

- I -

IMAGINATION/IMAGING. This is a powerful technique of being receptive to or actually attaining the good in our lives. Imaging is a total mental focus and concentration on a good we wish to be realized in our lives. Imaging is an advanced attention and expectation beyond imagination. This proven technique is an integral part of this program.

INFLATABLE CHUTE; INFLATABLE SLIDE. A device that extends from the door of an airplane to a surface. In the event of an emergency situation passengers exit the airplane quickly by sliding down this device.

INSTRUMENT LANDING SYSTEM (ILS) A generic term indicating that pilots may select one, or some combination, of several high-tech systems to guide the airplane accurately to the landing end of the active runway at the destination airport.

INTERNAL POWER. When the cabin lights "blink" in the airliner at the gate and prior to departure, the power for electricity, air conditioning, and heating is now within the airplane because the engines are started and are producing internal power.

- J -

JET ENGINES. With the invention of the jet engine, which is used on all large commercial jet airliners, the aviation industry was dramatically revolutionized. The jet engine is remarkably simple in operation and fantastically dependable, and it produces tremendous power.

JETWAY. The mechanical walkway at a gate connecting the waiting area to the airplane. Passengers at larger airports enter and exit airplanes via the jetway. The jetway is self-propelled and is moved by a steering apparatus up to the airplane, after the airplane is fully stopped.

- L -

LANDING GEAR. The mechanism of airplanes containing the wheels and tires. Landing gears retract into the airplane after take-off; they are extended or lowered prior to landing.

LAP BELT. The device that wraps around a passenger while seated in the airplane. Also called "seat belt." Federal Aviation Rules require that lap belts be fastened on take-off and landing, and at all times when the "fasten seat belt" sign is illuminated. Most passengers feel especially secure when their lap belt is fastened.

LAVATORY. The name of toilet and restroom on an airplane.

LIFT. One of the four aerodynamic forces, fundamental forces, physical laws, operating on an airplane.

Lift is primarily generated by the flow of air around the wing. Lift must be greater than weight in order for an airplane to fly. See also AERODYNAMIC FORCES and WEIGHT.

LIGHTNING. Lightning near or striking an airplane is frightening to many passengers. Because the airplane is not grounded, any damage is extremely rare.

LIGHTS, AIRPLANE, RUNWAY, TAXIWAY. Landing lights are always turned on when an airplane is in the vicinity of an airport. Lights outlining a runway are always white except at the ends: at the landing end they are green, and at the opposite end they are red. The lights outlining taxiways are always blue.

- M -

MEDICATION. Medications are sometimes prescribed for anxiety, they may be used either alone or in conjunction with other types of treatment/therapy for severe anxiety. This program uses behavior therapy and cognitive therapy.

MEDITATION. One of the techniques suggested in this program to reduce anxiety and fear.

MIND/BODY/SPIRIT. Very, very clearly, this program combines all three. This is one of the several qualities that makes this book/program so creative, unique, and effective.

- N -

NOISE LEVEL. A passenger who is anxious or afraid has a heightened sensitivity, particularly to noise changes. This program lets the user know in advance when noise changes will occur and the meaning of those changes in noise.

- O -

OUT-PLANE EXPERIENCE. Passengers who feel comfortable doing so may look out the airplane's windows while in flight and receive fantastic experiences from cloud formations.

OVERHEAD COMPARTMENT. Compartments at the top of the cabin on both sides of the plane. Passengers may store their carry-on luggage there.

OVERHEAD READING LAMP. Each passenger has a light over her or his head. The passenger decides whether or not to turn on this light. The light may be used all during the flight if the passenger desires.

OXYGEN MASK. In the event of an interruption of oxygen coming into the cabin, a small door over the head of each passenger will open and an oxygen mask will drop down in front of each passenger. Passengers simply put on the oxygen mask, as the flight attendant has demonstrated how to do before the flight began.

- P -

PHYSICAL LAWS. The airplane, its operation and flight characteristics, and the air, are all governed by physical laws. These physical laws are predictable and consistent. There is no mystery, no unknown, regarding the airplane and flight.

PHYSIOLOGICAL REACTIONS (SYMPTOMS). There are definite physical reactions, also know as "symptoms," to anxiety and fear. The physical reactions or symptoms, however, are caused by emotional rather than physical problems.

PILOT IN COMMAND (PIC). The captain is always the pilot in command. The captain possesses ultimate responsibility and authority for everything with the airplane and the flight. He or she makes all final decisions.

POWER, AIRPLANE. The commercial jet airliner is a powerful machine. Airliners have at least two engines. The airplane could fly on one engine. There is more power in a jet airliner than is hardly ever used.

POWER, SELF/SPIRITUAL. This program is designed to bring into reality the tremendous power that exists in human beings. At the same time, there is a clear encouragement to draw upon the Power that is beyond us and still a part of us.

PREDEPARTURE/PRELANDING ANNOUNCEMENTS and DEMONSTRATIONS. FAR's require the flight attendants to make specific announcement and to perform specific demonstrations. These are important, and all passengers should pay close attention when these are being made.

PREFLIGHT. Also known as "visual inspection." One of the pilots, or the flight engineer, <u>must</u> walk around the outside of the airplane and look carefully at, and touch if appropriate, the various parts of the airplane <u>before</u> <u>each</u> <u>flight</u>.

PREMOVING OPPORTUNITY. The ideal time for a passenger who is anxious or afraid to apply and practice the techniques and approaches of this program is before the airplane has moved away from the gate. The purpose is to feel—as much as possible—relaxed, confident, and secure and to be eager and excited with anticipation for the flight to begin.

PSYCHOLOGICAL REACTIONS. Strange things happen to the way our physical and mental sensors function when we are anxious or afraid. Our feelings and emotions and perceptions do not give us accurate feedback or input when we are anxious or afraid. Our feelings and emotions and perceptions, all this time, are not dependable.

- R -

RECEPTIVE VOID. One technique in meditation during which a person is receptive to whatever spiritual truth may be received. This is sometimes referred to as "silence," or "going into the silence."

RELAXATION. A vital ingredient in this program. Mental focuses and/or music assist greatly in relaxation.

RUDDER. A mechanical small wing-like part located on the vertical stabilizer, the big fin sticking up on the tail section, manipulated by the pilots on the flight deck, to turn the airplane around the yaw axis.

RUNWAY. A clearly defined rectangular area on a land airport that may have a hard surface, such as cement or asphalt, or a soft surface, such as grass. The runway is prepared and used exclusively for take-off and landing aircraft along its length.

- S -

SECURITY CHECKPOINT. Located at the entrance to every concourse and terminal building through which every person must pass to the waiting area and gate. You and all your luggage, and everything you are carrying, wearing, or have on you in any way will be carefully checked by some kind of electronic device. A personal search of your luggage or person may occur. Do not make any jokes that might even hint of foul play or act in a suspicious way, in fun or otherwise, at this location.

SENSES. There are two particular times and places when traveling by commercial jet airliner seems to cause some of our senses to "play tricks" on us. One is when the airplane is slowly backing away from the gate. The other is when an airplane is turning slowly in flight, especially when in a holding pattern. Our senses probably give us the most exciting "message" when an airplane begins acceleration, rotation, and lift-off.

SMOKING. It is usually against the law to smoke in an airplane on short domestic flights. Smoking should never be done in the lavatory.

SMOKING/SEAT BELT SIGN. A sign clearly visible to every passenger, controlled by the pilots, that indicates when smoking is and is not permitted and when seat belts must be fastened. Passengers must follow the directions of the sign. Note that a bell sound is made at the same time there is a change in the sign.

STABLE. An airplane is designed to be stable, and in fact it is stable. When an airplane appears to be unstable to passengers, the airplane is simply following the characteristics and movement of air through which the airplane is flying. The airplane never ceases to be other than stable; the air through which the airplane is traveling may be unstable.

STORM. Every commercial airliner has radar on the flight deck. This allows pilots to see a storm long before the airplane arrives at that storm. Pilots request and receive permission to deviate from their flight plan in order to fly around a storm.

- T -

TAKE-OFF. The beginning and accelerating movement of an airplane on the active runway. As the airplane attains a predetermined speed, the airplane is literally "tipped back" by the pilots into the most efficient configuration to produce maximum lift. This position called "rotation" will be maintained until the airplane attains full flight capability and lifts off the runway and begins a climb to cruise altitude. This is one of the most thrilling parts of flying by airplane to many people.

TAXIING. The aviation term for the movement of the airplane under its own power between the gate and the edge of the active runway. An airplane is never "driven"; an airplane is always "taxied."

TAXIWAYS. The designated area on an airport used exclusively for the movement of airplanes.

TERMINAL BUILDING. Probably not the best name for the main building containing airline personnel and passengers services for flying on a commercial jet airliner.

THRUST. An aerodynamic force, or a physical law, or a fundamental force, that provides forward movement of the airplane mostly because of the tremendous power generated by the jet engines that overpowers the force of drag.

TRAFFIC ALERT ANTICOLLISION AVOIDANT SYSTEM (TCAS). A very sophisticated high-tech avionic that informs pilots that another airplane is in, or is likely to be in, their air space. When two or more airplanes even begin to approach the same air space, this avionic in all the airplanes involved becomes self-activated. Information is given to all pilots by a word display and by a voice. This information instructs all pilots what action, if any, is needed.

TRAVELING COMPANION. A nice addition to the contents of this program is to make a flight accompanied by a companion. Note that the companion should be familiar with the program and be supportive of working the program <u>exactly</u> as the program is written and presented for use.

TRAY. A folding table attached to the seat in front of each passenger. This table offers numerous uses to the passenger. The FAA requires that the table be folded up securely during take-off and landing.

- V -

VENTILATION VALVE. Each passenger location contains an overhead ventilation valve. Both the amount of air and the direction of that air flow may be controlled by each passenger. Persons with anxiety feelings might consider directing a substantial flow of air onto them.

- W -

WAITING AREA. A designated area at each airline for passengers who are waiting to board an airplane. At many airports—and largely determined by the international situation—visitors may be with departing passengers or meet arriving passengers.

WEIGHT. Weight is one of the aerodynamic forces, fundamental forces, or physical laws that must be overcome in order for the airplane to fly. The force of lift must be greater than the total weight of the airplane.

WIND. Wind is air moving. Most wind is caused by air flowing from a high pressure area into a low pressure area. Wind is a very important factor to pilots and flight; pilots receive accurate information regarding all weather, including wind information, before and during the flight.

WINDOW SEAT. A window seat, especially one over the wing, allows the user of this program to identify and confirm the operating parts on the wing of ailerons and flaps and dive brakes. The excellent unobstructed view also helps the user to experience an incredible view.

WING. The wing is the largest attachment to the fuselage of an airplane. Wings are designed so that air flowing around the wing produces lift.

INDEX

The first page figure listed after each subject refers to the most prominent presentation or coverage of that subject.

YOUR RESPONSE IS WELCOMED

CUT ALONG THIS LINE PLEASE

The specific design and intention of this program were to reduce your anxiety about flying and to increase your enjoyment of flying on a commercial jet airliner.

We would deeply appreciate your letting us know how you feel about flying now that you have completed the program.

I RATE MY FEELING AND THOUGHT ABOUT FLYING NOW WITH AN "X" ON THIS RESPONSE:

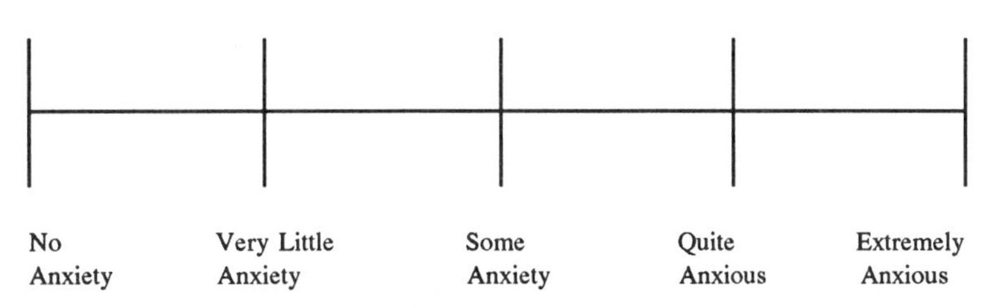

PRIOR TO USING THIS PROGRAM, I WOULD HAVE RATED MY FEELING AND THOUGHT ABOUT FLYING WITH AN "X" ON THIS RANGE:

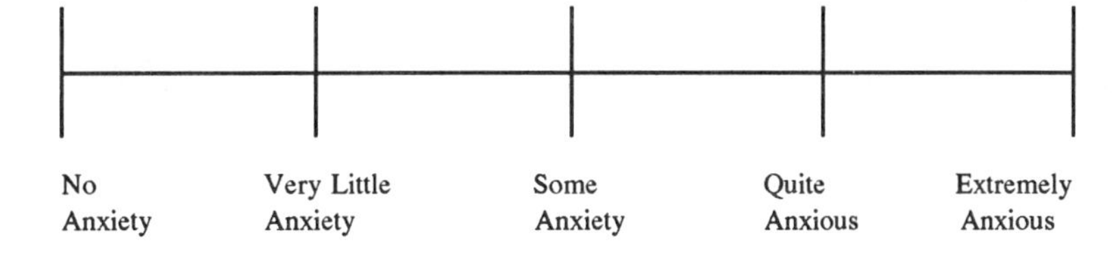

CUT ALONG THIS LINE PLEASE

COMMENTS ABOUT ME AND MY USE OF THIS PROGRAM:

___

___

___

___

___

___

___

___

Thanks so very much! You may or may not sign your name; you may or may not give your address.

Please mail to: Inner Marker To Growth
P.O. Box 23310
St. Louis, MO 63156, U.S.A.

BLANK PAGE FOR NOTES